Bikram's Beginning Yoga Class

Bikram's Beginning Yoga Class

Bikram Choudhury

with Bonnie Jones Reynolds

Second Edition edited

by Julian Goldstein

Photographs by Biswanath "Bisu" Ghosh

thorsons

Every effort has been made to ensure that the information contained in this book is complete and accurate. However, neither the publisher nor the author is engaged in rendering professional advice or services to the individual reader. The ideas, procedures, and suggestions contained in this book are not intended as a substitute for consulting with your physician. All matters regarding your health require medical supervision. Neither the author nor the publisher shall be liable or responsible for any loss, injury, or damage allegedly arising from any information or suggestion in this book.

Thorsons
An Imprint of HarperCollins*Publishers*
77–85 Fulham Palace Road,
Hammersmith, London W6 8JB

The website address is: www.thorsonselement.com

thorsons™

and *Thorsons* are trademarks of
HarperCollins*Publishers* Ltd

First published by Penguin Putnam Inc. 2000
First published in the UK by Thorsons 2003

10 9 8

© Bikram Choudhury 2000

Bikram Choudhury asserts the moral right to be identified
as the author of this work

Photographs on pp. 2, 4, 46 (top), 47 (top), 48 (top), 54,
98, 100–103, 116, 118, 164 (top), 165 (top), 166 (top),
169 (left), 194, 196, and 197 by Guy Webster.

A catalogue record of this book is available from the
British Library

ISBN-13 978-0-00-715499-9
ISBN-10 0-00-715499-2

Printed and bound in Great Britain by
Martins the Printers Ltd, Berwick upon Tweed

Acknowledgments

I wish to thank all my students who have proven through their dedication and practice what we have known for centuries in India about the curative and restorative effects of Hatha Yoga. In the years that followed the building of my first school, and the opening of Bikram's Yoga Colleges of India around the world by my certified teachers, the legacy of my Guru, Bishnu Ghosh, is being fulfilled. To all those who have and are helping me fulfill this legacy, I give my heartfelt thanks. I want to give special recognition to Emmy Cleaves, my most senior teacher, and to my wife, Rajashree Choudhury, for running the day-to-day activities of the Yoga College of India and her contribution to my teacher training program, without which, the program would still only be an idea. A special thanks to all my students who gave their time to pose for the pictures in this book, and a very special expression of gratitude to my lifelong friend and the son of my Guru, Biswanath "Bisu" Ghosh, who is responsible for the shooting and development of the photos for this revised edition. And thanks to my friend, Julian Goldstein, without whose help this second edition would not have been completed.

For My Guru, "Yogindra," Bishnu Ghosh

Paramahansa Yogananda, founder of the Self-Realization Fellowship, brother of Bishnu Ghosh

Contents

Before undertaking any posture in this book, it is imperative that you read the Introduction and the posture instructions thoroughly.

Introduction

We've Come a Long Way

In 1978 I finished this book, **Bikram's Beginning Yoga Class.** I told my publisher that this book will sell forever like a bible. He laughed and asked, "Why do you think that this book will do so well?" I told him that new people will always come, buy the book, and do Yoga, especially as Yoga becomes more popular day by day. And it has. And this book has sold very well and it still is popular among people who are seeking the healing remedies of true Hatha Yoga. Today it is more popular than ever. In 1978, if you mentioned the word Yoga, no one even knew the meaning of the word. Today everyone knows something about Yoga, even if it's only knowing the word itself. Also, the popularity of this book has spawned my Yoga Teacher Training Program at my Yoga College of India. Today, there are yoga schools all around the world with my Certified Teachers teaching my Beginning Yoga Class.

Why Did You Pick Up This Book?

Why did you pick up this book? What is your problem? What are you looking for? Maybe you have a chronic disease like heart disease, diabetes, hypertension, kidney disease, a glandular disorder, Parkinson's, or another neurological disorder. You could have respiratory, digestive, or spinal problems, or arthritis. There are too many things that go wrong to list here. Perhaps you're not happy. You are yelling at the kids, angry all the time, depressed, anxious, or afraid. You're overweight and every time you try to lose weight you end up gaining more. You seek peace for yourself in the chaotic world you live in. You have problems and you don't know what to do. You've tried medication and found that it doesn't work. First you take the medicine for your problem, then you develop another problem because the medicine you take causes side effects. Now you have two problems and your medicine chest is filling

up fast. My guru said that stress and strain are the causes of all chronic diseases, even infectious ones. My Beginning Yoga Class is the great stress-buster.

What about surgery? Suppose you have an operation for a herniated disc. You solve five percent of the problem, but then you create a new problem. The surgery often takes more away from you than the problem you had to begin with. Surgery doesn't work. Still you are looking to solve your problem. In India we know that where medical science stops, Yoga science begins.

Maybe you don't think you're so bad off. Just a little out of shape. Maybe worse. You have no energy and you can't climb stairs without huffing and puffing. You've tried to exercise—aerobics, swimming, weight training, stair climbing. You've played tennis and many other sports. You've used machines for your legs, arms, back, and abdomen, and still nothing works. When you do sports you are hammering your body in a cold environment and you harm the body. First you destroy your skeletal system, then you destroy your nervous system, then you destroy all the tendons and the ligaments, then the veins and the muscles. Everything you have tried fails you. You are not the failure, your system is. Exercise doesn't work. When you do your exercise you only exercise three percent to ten percent of your body, besides the damage that you do. When you do my Beginning Yoga Class, you exercise one hundred percent of your body from the bone to the skin, from the head to the toes, to every gland and organ of the body, to every cell, even to tiny tissues. You have been looking in the wrong places. Now with this book, you are looking in the right place.

What you are looking for is Yoga. You are looking for my Beginning Yoga Class. Yoga is the only exercise on this planet from which you gain energy instead of burning energy. That's why my guru said, "Yoga maintains youth long. It keeps the body full of vitality, immune to diseases, even at old, old age. The yogi never becomes old. Yogis achieve the supernatural power."

How It All Started

In this book, you will learn the Hatha Yoga asanas (postures) as set down by Patanjali over 4,000 years ago. This Hatha Yoga is for everyone and every body. It doesn't matter how well you do each posture, only that you **try the right way.** Even if you can only do part of the posture, you will receive one hundred percent of the benefit medically if you **try the right way.** I explain this by giving you step-by-step instructions for each posture.

Before I wrote this book, I had already been teaching Yoga for more than fifteen years. Also, I had already developed the twenty-six postures and two breathing exercises which have come to be known as Bikram's Beginning Yoga Class. I discovered and developed this scientific series of poses and breathing exercises during years of research. This series has been proven over and over again to cure chronic diseases of all kinds. I have confirmed the value of this Yoga series again and again, scientifically, medically. In the years that followed the publication of this book, millions of people around the world have used it. They followed my instructions and they cured and healed their problems—medical problems, emotional problems, and chronic diseases. Today, I know how good this book really is because I have seen the results with my own eyes. The important thing that this book brings you is instructions on how to do my twenty-six postures and two breathing exercises, exactly.

In order to understand this book in its entirety you have to open your heart, mind, soul, and eyes—open your everything, your every sense. If you really want to learn and gain an understanding of human life, your life, and understand how you can change your life, you must accept the principles you will find in this book. If you think you are American, Chinese, Japanese, Indian, Christian, Jewish, Hindu, or Moslem, if you think you are white or black, rich or poor, then you are not going to get any benefit from the philosophy of this book. You will still get the benefit from practicing the Hatha Yoga, in any case. However, if with your open mind you can accept the principles in this book, your attitude will slowly change about who you are. Your relationship with all humanity will change. You will begin to realize that each one of us is one of six billion people, a part of the world society and all civilizations. We must think this way because we are not alone in this world. We are not divided up by boundaries, borders, or cultural backgrounds. The best of all cultures is available for each of us. But we must know them first. That's why you must be open-minded to get the greatest benefit from this book.

What Is Bikram's Beginning Yoga Class?

First, I want to tell you that Hatha Yoga is not mine. What has become today known as Hatha Yoga is thousands of years old, older than the history of earth and mankind that we study in school. For thousands of years the discipline has been handed down from master to disciple, one to one. My first training of Hatha Yoga came from my brothers' teachers when I was three years old. Soon after my initial training, I met my guru, Bishnu Ghosh, from whom I received my formal training. My guru was the younger brother of Paramahansa Yogananda, the great spiritual leader. He was Yogananda's first disciple. My guru was the greatest physical culturist in the twentieth century. He scientifically established the methods of curing chronic disease using Hatha Yoga. Before my guru sent me to Mumbai (Bombay), I had established myself by winning important Hatha Yoga competitions and excelling as a runner and weightlifter. I did these things at my guru's urging. He wanted to demonstrate to the world how Hatha Yoga improves the human condition mentally and physically.

When my guru sent me to Mumbai to teach Hatha Yoga to sick people, I found that there were more people who needed me than I could help. There wasn't enough time in the day to help everyone. I thought that if there was some way I could teach everyone the right postures in exactly the correct order, no matter what their disease condition was, then I could teach people in groups and then help more people. This was a totally different approach than the one-to-one, master-to-disciple method that had been done for thousands of years.

I researched the diseases and the postures and after many years of research and verification, having used the methods taught to me by my guru and using modern medical measurement techniques, I arrived

at the sequence of postures you will find in this book. The significance of this series is that no matter what condition you are in, what chronic disease you may have, or how old you are, the solution to your physical and mental problems is in doing these twenty-six postures and two breathing exercises. That's why I always say, "It's never too late, it's never too bad, you're never too old, you're never too sick to start from scratch once again, to be born once again."

Warning

Unfortunately, Hatha Yoga has been badly abused in the West. A number of yogis came to the West to teach Hatha Yoga. They and their disciples destroyed the Hatha Yoga System as it has been known for thousands of years. These yogis knew true Hatha Yoga, but because of their lack of faith in the Hatha Yoga system and the Western people, they have ruined Hatha Yoga in the West. Ruined it even down to the three basic principles of freezing the body in the posture for the prescribed number of seconds, proper control of the breath while in the posture according to the posture, and following the posture with the minimum of twenty seconds in savasana, complete relaxation.

These yogis and their disciples failed to recognize that because of the cultural differences between the East and the West, teaching methods that worked in the East could not work in the West. Rather than change the method by which they taught, they changed what they knew to be true into something that merely resembles Hatha Yoga, but does not deliver the goods. The damage that these yogis did by not holding to Patanjali's **Yoga Sutras** has caused damage far beyond those whom they taught. They led their disciples to believe that they had learned Hatha Yoga. Of course, they had not. To compound the problem further, their disciples were taught with a total lack of regard to the Hatha Yoga system, which is grounded in discipline. Then these disciples felt that they could do the same as their teachers.

Americans are very inventive, even if what they invent is wrong. They are inventing posture after posture, making up names for them. Then they sell their wares to innocent, uninitiated people who do not know that they are getting ripped off, even getting hurt. There are more flavors of Hatha Yoga in the West than ice cream. Americans think that's wonderful. I tell you it is disastrous. Most of these so-called Yoga systems are not Yoga at all. Giving them a Sanskrit or Bengali name doesn't make them Yoga. And using props to help you do the postures only makes matters worse and not better.

If you attend a Yoga class, how do you know if the instructor is qualified? Just because he or she says so is not enough. The teachers I train go through a program that is equivalent in time and study to a one-year university professional-degree program. My teachers go through rigorous physical, mental, and spiritual training in my Yoga College of India. Before they receive a certificate they must demonstrate their knowledge in an actual teaching environment. My teachers know the medical benefits of each posture. Most Yoga instruc-

tors do not. My teachers know how to direct you to **try the right way,** the very foundation of your Hatha Yoga practice.

When you attend my Beginning Hatha Yoga Class, the first thing you will notice is the room is heated. If you do your Hatha Yoga in an air-conditioned room, you can easily hurt yourself. I explain that more in detail in the next section, **Getting Started.** There are the three requirements that I mentioned before, freezing the body in the posture for the prescribed number of seconds, proper control of the breath while in the posture according to the posture, and following the posture with the minimum of twenty seconds in savasana, complete relaxation. If you do not practice in this manner, you do not receive the medical benefit of the posture. This book is your guide for the proper sequence of postures and the ideal position for each posture. There are no alternatives. There can be no substitute for true Hatha Yoga.

Getting Started

Before you start there are a few things that you will need to help you do your Yoga practice. Also, there are a few things that you should know to help you approach the postures so that for each pose you attempt, however little or much you can do, you will always be **trying the right way.** Thereby you will always be receiving one hundred percent of the medical benefit, no matter what your physical or mental condition.

Where to Do Your Yoga

You need a place to do your Hatha Yoga. You don't need very much room. Even if you live in a tiny one-room apartment you'll find enough room. Basically, you need a clear area of floor where you can lie down on the floor and stretch your arms sideways and lie with your legs fully extended.

Special Tools

You need a space on the wall to hang a mirror. This should be a full-length mirror.

You can wear just about anything you like as long as it fits tightly to your body. The important thing is that the clothes you wear do not bind when you do the postures, and you can see the position of all your body parts as you observe yourself in the mirror.

You will also need something to spread on the floor. If you purchase a sticky Yoga mat, spread a towel over it so you receive full advantage of the postures. If your floor is not carpeted, you will need something underneath your towel so it won't slip on the floor.

Make It Hot

You must heat the area where you do your Yoga. If you can, you should try to heat it to at least 100°F. You should sweat a lot when you do your Hatha Yoga. If your bathroom is large enough, you can preheat the room with a space heater and by running the shower with hot water. Leave the water in the tub as this will help keep the bathroom hot.

If you have difficulty heating an area to 100°, then you must wear warmup clothing

while you do your Yoga. This will keep the heat from escaping your body.

I cannot overstate the importance of doing your Hatha Yoga in heat. Doing your Yoga in a cold environment can bring harm to your body. Remember, you are changing the construction of your body as you perform these postures. Suppose you are going to make a sword. You start with a piece of fine steel and the first thing you do is put the steel in the fire and heat it up. When the steel is hot, it becomes soft. Then you can hammer it and slowly you make it change shape to the sword you want. This is the natural way. Now if you don't heat it up and start hammering the cold steel, nothing is going to happen to the steel, but you'll break your hand, the hammer, your arm, and all your connecting joints. The same thing happens when you do any exercise, even Hatha Yoga, in a cold environment. When you do your Hatha Yoga in the heat, your body is malleable.

A Word About Balance

In the instructions for the Yoga postures that follow, I will not give you instructions on how to balance. Although you may find it difficult in the beginning to balance in some of the standing postures, you will find that with continued practice of these postures, your ability to balance will improve. You should not use a wall to support yourself, even in postures that you think make it impossible to maintain your balance. It is better if you try the posture on your own steam even if you can only hold your balance for half a second to begin with. That half-second will become one second, that one second will become two, and so on as you continue your practice, until finally you can hold the posture for the full time.

Try the Right Way

I cannot overemphasize the importance of **trying the right way.** It is how you try that is the single most important factor in the benefits you receive. The fact that you cannot do the postures perfectly, according to the directions in the book and the beautiful pictures of my students, should be of no consequence to you. These directions and the pictures are your ultimate goals for the postures. If you try hard today, apply your concentration one hundred percent, follow the directions step by step, and use your strength where called for, then you are doing a perfect posture for you today. This is what it means to **try the right way.** You give a one hundred percent effort in all aspects, and you receive one hundred percent of the benefit, medically, physically, and mentally.

Stay With the Book

Don't jump around. The order in which you do the postures is as important as how you do each posture. The sequence **must** not be altered. The series of postures requires that you must do the first posture before you do the second posture. The third posture follows only after you have done the first two, and the fourth after you have done the first three, and so on until you have completed all twenty-six postures and two breathing exercises.

Precautions

You should not eat before doing your Yoga unless you want an upset stomach. Three hours before Yoga is a good rule to follow. Take your meals after your Yoga practice. Also, if you are trying to lose weight, you will find that you are not as hungry after you do your Yoga.

Hatha Yoga is not aerobics, so when you start a posture or you come out of a posture you must do so very slowly. Remember, you are changing the construction of your body. Also, when you are making corrections toward achieving a perfect posture you should make tiny changes. You should treat how you learn these postures as if you are a baby and learning how to walk. You are taking baby steps.

Breathing

Each posture has a particular way that constitutes correct breathing for that posture. I call that the **normal** breathing for the posture.

First, there is *80–20* breathing. In this method of breathing, you take in a full breath. Go to the posture and continuously let out 20 percent of the air through your nose with your mouth closed. In postures that require *80–20 breathing,* you will need oxygen in the lungs to do the posture, so you will be able to maintain proper strength while performing the posture.

Second, there is exhalation of the breath. In *exhale* breathing, you take in a full breath and exhale the breath completely when you achieve the posture. While you are doing the posture you should continue exhaling.

With either breathing technique, you should not strain. In the beginning, your lung capacity will not be large enough for you to sustain yourself with the above breathing methods. To prevent straining your lungs, take another breath as needed and continue with the breathing method. As your lung capacity improves and you improve doing your Yoga postures, you will find that following these breathing methods becomes as natural as doing the Yoga postures themselves.

You are now ready to go. So turn the page and begin. . . .

PRANAYAMA SERIES
Standing Deep Breathing

Janice Lynde

ONE

The room seems unreasonably hot. You hesitate in the door of the dressing room, clutching your towel. You work out your strategy. This Choudhury fellow that all your friends have been raving about is sitting on a mound of pillows at the front of the room. Several students are lounging about on that little island with him, talking, laughing, and helping themselves to his array of candy, cookies, fruit, and nuts. He hasn't noticed you yet.

And pray God he won't. You tiptoe past knots of chatting students to the back of the room, position yourself behind a big, tall, blond fellow, and wait. Towels dot the smooth, clean, wall-to-wall carpeting, towels dropped by their owners to stake out their favorite positions.

The heat, you will find, loosens and relaxes the muscles, allows you to do the exercises more easily and, thus, earn the approval of Bikram—and yourself.

For the moment, however, you are sweating and irritable, wondering why you've come. You are also acutely nervous and embarrassed, because everyone is watching you. The very fact that they are all acting as though they aren't aware of you is proof of that fact. You pluck with discomfort at your leotard/bathing trunks. (You tug your bra down and tuck your panties back up out of sight if you are wearing a leotard—which means you are female—and

observe with shock and censure that most of the other females are wearing absolutely nothing under their leotards. If you are in bathing trunks, which means you are a fellow, you might feel something besides censure at the observation.)

You move more squarely behind that big fellow in front of you so that you don't have to see yourself in the full-length mirrors ranged across the front of the room and thus distress yourself still further. Because, in your leotard/bathing trunks, according to actuary tables and regardless of your age, you **are** a distressing sight. The bulges normally hidden beneath street clothes are clearly and painfully revealed. So are your flabby muscles. You sneak another peek at this "Choodree" fellow, noticing that **his** muscles look like Michelangelo's anatomical sketches.

So, you realize as you look around, do the muscles of some of your fellow students. If it weren't for the soothing presence of a ten-year-old girl, a wizened old sprite, and several comfortingly pillowed bodies scattered around the room, you'd swear you had blundered into an advanced class. With that, you see long, lithe creatures in the front row, limbering up with movements that would frighten a pretzel—and you **know** you've blundered into an advanced class. Horrified, you attempt to organize your flabby muscles and bolt the room, at which

point Choudhury springs to his feet and claps his hands.

"Okay. We start now."

You freeze. Around you, conversation ceases and the students glide to their favorite positions.

"Who we got is new today?" Choudhury's eyes scan the class. He must have X-ray vision. "Hey. Who is that person hiding behind Reggie? Why you hiding, Person? Come out where I can see. . . . Oooooooohhh!" (It sounds as though he has just found his pet dog poisoned.) "My goodness, I can see why you must hide, look that junk body. Good thing you come to me, I fix you up. Come up here. Comecomecome, in second row, nice in middle behind Shirley and Archie so you can see yourself in the mirror. Seeing in mirror is very important for beginner. Can you see yourself? Good. You just watch Shirley and do it just like her.

"What's your name? Terry? Hi, Terry, nice to meet you, I'm Bikram. But I already got one Terry, so I got to call you Terry Two, okay? You ever do Yoga before? Well, don't worry, in one week coming every day you be one-hundred percent, just like all my students. There are just a few things you should know before we start. I will explain them more later, and say them over and over during the class till you begin to think I am a robot with a broken record in my belly. But, funny thing, no matter how many times I say them, half my students do not hear."

"That's because our senses are deadened from all your chatter," says a forty-nineish lady with a girlishly trim figure, a bouffant hairdo, and a sardonic mouth. "Maybe you should try quality instead of quantity."

"No. I just got to work harder to tighten up all the loose screws in your head, Florette. They still rattling so much. Every day this week, Terry Two, I told Florette not to come into my class wearing stockings again, and look at her. Beginners cannot do the exercises properly without bare legs and feet. Only very advanced students allowed to break my rule. You come in your stockings again tomorrow, Florette, I will not allow you to take class. This is no beauty contest, no one look at anybody else here.

We don't care you got varicose veins or cottage cheese on your thighs. We only care you do the exercises.

"So, Terry Two, that is the first thing. You got to listen with your all three ears and do what I tell you **exactly**. But you won't. I will tell you to **go into the pose slowly**, and you will rush. I will tell you to hold the pose **honestly** and exactly with your all strength until I say to stop, and you will cheat. I will tell you to **come out of the pose slowly,** not collapse like sack of wheat, and you will collapse. I will tell you to **breathe normally** while you are in the pose. Then I will tell you again to breathe normally, and still you will hold your breath. I will tell you **not to close your eyes** at any time during my class, and you will close them. I will tell you to **relax** so many times I will lose my voice, and still you will be all tight and fight the posture. I will tell you **concentrate on one spot** and **hold the pose like a statue,** and you will bob around like toy duck in ocean. . . ."

"Gee, Bikram, why do you bother to tell us any of these things if you know what we'll do?"

"I donno, Archie, I'm crazy, I guess. I am searching for a student who will finally listen, just like that Greek who wandered around with a lamp looking for an honest woman."

"I think he was looking for an honest man."

"Oh? Well, that's okay, too. But the most important thing I will tell you, Terry Two, is that as a beginner you must do Yoga every day. All these other things, like going in and out of the poses slowly, giving it your full honest effort, breathing normally, keeping your eyes open all the time even during rest periods, relaxing, concentrating, holding the pose like statue . . . I will forgive little bit if I got to say it a few thousand times before you hear me. But I won't forgive you if you don't come every day to class for at least two months. Only on Sunday you don't need to do Yoga. Only crazy people do Yoga on Sundays. How many crazy people I got in this class?"

Almost all the hands raise.

"See? Nobody listens to me," he moans.

"Tell Terry Two about The Cumulative, Bikram."

Bikram turns warm brown eyes to the ten-year-old, a lithe little blond. "Why don't you tell it, Barbie?"

Barbie reddens. "I'm not very good at arithmetic."

"Neither is Bikram," says Florette. "The Cumulative is the craziest contortion of figures I ever heard."

"Then how come it works?" asks Barbie.

"Florette," says Bikram, "how'd you get so mean with such a nice, flowery name?"

"Technically," says Archie, "even a Venus Flytrap is a flower."

"Hey you guys," says Shirley, "I don't know whether the rest of you have to earn a living, but I'm supposed to be at a rehearsal in two hours."

"Okay. I'll tell you about The Cumulative later, Terry Two. Just one more important thing I got to tell you before we start. There is nothing I ask you to do in this class that is going to hurt or injure you. Just follow my instructions **exactly,** you will be safe. Even if you are convinced you are going to break in half or your head will fall off because you forgot to screw it on, it won't happen. All beginning students are scared, and it is biggest obstacle you must overcome to make progress. Don't fright, don't scare, I will take care of you. Okay? You do just as much as you can honestly do the first day. You don't have to be hero. The best you can do is all that I ask. That is perfection in Yoga, the best you can honestly do on any particular day.

"So, we begin with a breathing exercise. This is to increase the capacity and elasticity of your lungs and move oxygen into your bloodstream, get it to all the corners of your body so they are awake and ready for the other exercises. Any time, any place you are tired and need energy, do this breathing exercise and your vitality will come back. Watch now, I will show you."

1

Stand with feet together, pointed directly at mirror. Interlace your fingers. Lift your hands, palms together, and place the knuckles of those interlaced fingers firmly under your chin, and keep your elbows together. You will keep this knuckle-to-chin contact at all times during the exercise.

IDEAL
Pranayama

1

The first thing you will notice about the ridiculously simple maneuver of standing with your feet together is that you can't do it. Feet together means heels together, bunion bones touching. Once you get this firmly in mind and keep reminding yourself, your heels will behave most of the time, but your toes won't. They will refuse to stay together and pointed directly forward because you will feel as though you're falling to one side or another. You'll be tempted to spread your toes for support. But keep them together and don't panic. You won't fall over. And you will soon feel comfortable in the stance.

From here on, whenever I direct you to put the feet together, this is the stance I want you to assume.

Most people are able to interlace their fingers and put them under their chins without much trouble—but then things get tough. Avoid the common error of arching or breaking the wrists upward, leaving the

elbows dangling. Keep the wrists relaxed and down, with a nice straight line from knuckles to elbows.

REALITY
Standing Deep Breathing

2

Keep your mouth closed and inhale deeply by way of the nose, but actually breathe with the throat. In this breathing series the mouth and nose are only the passages by which air moves into or out of the lungs. Inhale slowly, steadily, as much air as possible for a long count of six. Feel as though your lungs are a glass of water that you are filling from the very bottom up to the rim. **Simultaneously** as you inhale to a long count of six, slowly raise your elbows like seagull wings on either side of your head (your goal is eventually to raise your elbows until your forearms touch your ears), and resisting with your fingers, lower your chin into the crotch of the "U" that your arms will form. Do not bend forward, just lower chin.

More

More

2

You'll be shocked to find that, at about three and a half, your breath will shudder to a strangulated halt. You'll stand there holding your breath for the next few counts feeling slightly foolish and worrying about your arms, which don't seem at all like seagull wings and are nowhere near touching your ears. Don't be discouraged if you can't raise your arms above shoulder level at first. Concentrate on pressing the chin against the knuckles so firmly that you finally make them crack, and that flexibility in the finger joints will turn you into a graceful gull before you know it.

As for the breathing, that will come when finally you understand what is meant by "breathing with the throat." If, on the intake, you feel it in your nostrils and make a sniffing sound, you're not using your throat. To get the air where it belongs, you must pull it in steadily through your nose, until the pressure of it forces a snoring sound in the back of your throat. (Actually, the first few times, it will be more than just a snoring sound. Take it in stride if a litter of piglets comes running, mistaking you for Mother. Things will quiet down once those throat muscles grow supple, I promise.)

3

Immediately and in a fluid movement once the count of six has been reached, open your mouth just a little bit and let the exhalation escape slowly and steadily out of the mouth for a long count of six, **simultaneously** dropping your head backward max-i-mum, keeping knuckles to chin, and bringing your arms, wrists, and elbows forward to meet in front of your face. Feel now that you are emptying the glass of water, forcing out every drop. Keep your fingers interlaced and your knuckles against your chin. Your elbows, wrists, arms, and up-turned face will then form one nice straight plane parallel to the ceiling.

IDEAL
Pranayama

3

The exhalation will be even more disconcerting than the inhalation. What a red, apoplexied face you'll have when the little air that went in refuses to come out. Could you possibly be that much out of shape? The answer is yes. (If you are a smoker who has been toying with the idea of quitting, this shocking look at your late, lamented lungs might give you just the jolt you need.)

To do it properly, just reverse the air flow described above, forcing it slowly and steadily up to ricochet off that same "snore" spot in the back of the throat under the nose, allowing the air to find its own way out of your slightly opened mouth. At the same time, be sure to touch the wrists as well as the elbows together.

You might also become dizzy or giddy at this point, for the sudden appearance of real oxygen in your tired blood can have the same effect on your brain as half a dozen champagne cocktails on an empty stomach. Be sure not to close your eyes at any point in the Standing Deep Breathing or you'll topple over.

REALITY
Standing Deep Breathing

4

Now do nine more inhale-exhale cycles, ten in all. At end of tenth cycle drop your arms naturally to your sides and rest a moment.

5

Do ten more cycles.

4

After a couple of inhalations and exhalations, you'll swear your arms had been poured full of lead. You'll begin to cheat by cocking the wrists and flapping your arms like waterwings. Soon even your hands will grow heavy and you'll have to fight to keep the knuckles up to the chin. By then your toes will have crept apart, your knees will have bent (not necessarily both in the same direction), and as you try to correct those problems you'll forget if your chin was supposed to be going down while the arms were going up and whether you were inhaling or exhaling and why. You will understand the meaning of eternity at last, for surely you've done many more than ten cycles, and yet the class just keeps going on and on.

5

It will seem like a hundred.

Benefits

Because of sedentary habits, most people use only ten percent of their lungs, never allowing the lungs to reach the maximum expansion capacity that Nature intended. As a result, they are susceptible to emphysema, asthma, shortness of breath, and dozens of other breathing problems. Standing Deep Breathing teaches you to use the other ninety percent of your lungs.

This exercise should be done before any kind of physical activity. Because it expands the lungs to their full capacity, it increases circulation to the whole body, waking everything up and preparing the muscles for action.

Classnotes from Florette

Welcome to Mr. Choudhury's Indian salt mine!

You know, it wasn't the hard work I minded on my first day. It was—well, would you believe to look at me now that I was forty pounds overweight? Lord! the guts it took for me to walk out in front of these people, half of them **men,** in a leotard. To add insult to injury, I felt clumsier than a baby elephant. I decided I wasn't coming back a second time, even though I felt like a million after the class.

Then, getting dressed the next morning, I found that the waistband of my skirt was actually the tiniest bit **loose.** So I decided I'd give this drill sergeant a chance. I'd look foolish for him an hour and a half a day if he could fix me so I'd look sensational for the other twenty two and a half hours. And he did. Now he's just impossible.

Shhh. He's ready for the Half Moon. What it does for the waistline!

ARDHA-CHANDRASANA
Half Moon Pose with
PADA-HASTASANA
Hands to Feet Pose

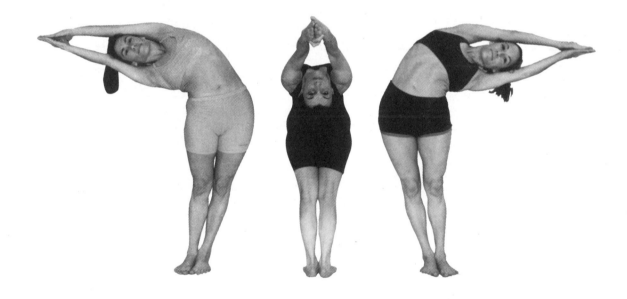

TWO

"The next pose is actually two separate poses, but since we do the Hands to Feet Pose with no break between, you can think of it as the fourth part of the Half Moon for when you practice at home." Bikram's eyes come to rest on a woman in the fourth row. She bears a remarkable resemblance to Humpty Dumpty. "Just like Lavinia, here, practice at home. 'Least she must, 'cause I never seen her in class. You practice at home, Lavinia?"

Lavinia maintains the visage of one well versed in ignoring what she does not wish to contemplate. "No. But I was here last Thursday. I come every Thursday."

"What you want, the class give applause? We give no applause for every Thursday. Every Thursday no good, Thursdays don't count. What about Friday and Monday and all those other nice days? What you got against them?"

"I'm too busy."

"You think my other students are retired millionaires? If they were I would charge what I ought to charge for each class—one ounce of pure, twenty-four-karat gold—and I be rich man quick and retire, too. Look Reggie here, he breaks stocks for a living and he got to get to office every day at six A.M. to make his ticker tape or whatever it does. But he been here every day for four months. Every day! Well, now I think, I remember he did miss three days. I think his dogs died

and he had to go to funerals or something. He got lots of dogs, we forgive that. How much weight you lost so far, Reggie?"

"Thirty-five pounds."

"That all? Looks like more. I think you also lose ten years. Lavinia here wants to lose thirty-five pounds and ten years, too, be photo of Dorian Gray like rest of my students. So she just now making resolution she going to come every day for two months . . . no. She shaking her head. What a stubborn lady, must be a Taurus."

"Thursday is the only day I can come. I have a family to take care of."

"Poor people, your family, do they know you don't love them? Oh. I can see that shocks you. You think you **do** love them, look how good care you take of them, you think—look how you sacrifice yourself. How old are you, Lavinia?"

"Thirty-five."

"Honey, I got to tell you, you look fifty. And your body acts sixty. Sorry, but you only going to get truth from me. I'm not here to make things easy for you and tell you what you want to hear. I'm here to save your life. You no time ever, Lavinia, if you live to be a hundred, gonna find anyone who loves you more truly than I do. If I didn't love you I wouldn't bother to be mean to you every Thursday. You letting your fat clog your heart and all your arteries; you letting your muscles turn to junk and your energy run away like through a sieve. Your joints are turning to concrete and soon you will have arthritis.

"Isn't that nice gift for your family? How happy they are gonna be with all the doctor bills they gonna have 'cause you are lazy and old before your time. Think how pleased your children gonna be when you get senile at sixty-five 'cause there's no way your blood can find your brain.

"You think this is love what you doing, Lavinia? Letting yourself become piece of junk? If you **really** want to take care of your family and show how much you love them, you first got to take care of **yourself** and love **yourself.** If you don't love yourself and take care of the body your god gave you, you will never be able to love anyone else, either. No, Lavinia, don't use your poor family as excuse for laziness. Greatest, most loving gift to family would be to get here every day for two months and get your body and mind young and supple and beautiful again. No, I'm wrong. **Most** greatest gift would be if you get your family to come here with you, like Charlotte here. See that man hiding in back row? He belong to Charlotte. He was very stubborn ninety-pound weakling, but finally he come to class with her and now he Tarzan and a happy man, right Charlie?"

Charlie, forty and trim, grins. "Anything you say, Bikram."

"You bet your boots, but I don't got to say. Every time you do Standing Head to Knee Pose and feel wonderful pain in your legs, I just got to look at your face to know you are **sublimely** happy."

"Bikram," asks Lavinia stiffly, "who wants to get into something they've got to do for an hour and a half every day for the rest of their lives?"

"Who told you that? Did **I** ever tell you that? Anyone who tell you got to do Yoga every day for rest of life, or that they do full set of Yoga every day for years, he crazy or saint or both. Regular people like us, we got to worry about the garbage disposal don't work, Tommy's scout troop has a picnic, there's a good movie on TV—important things like that. So some days we can't do. That's perfectly okay if you did your Yoga right to start with. Naturally, the more you do, the better health you gonna be in, mentally and physically. And if you got some

medical problem like bad back or arthritis or old age that Yoga is keeping away, you got to do the Yoga fairly regularly or you'll get it again.

"But the rest of us, once we got it, once the joints are opened, and muscles, ligaments, spine, sciatic nerves all stretched and supple and trained, we can relax little bit— do it only two, three times a week, or maybe just half a set when we're rushed. A half set, each exercise just once, takes about thirty minutes when you don't have to worry about me talking and gossiping in between. No one is so busy she or he can't exercise thirty minutes a day. And any amount of Yoga you do is going to give you some benefit.

"But Yoga is cumulative, you see, Terry Two. That's why I'm always mean to Lavinia here. She getting nowhere, she wasting her money. She come that first Thursday, it's like she get five points of worth in her body. If she came the next day she'd get five for that second day, too. And since the five from the first day wouldn't be wasted—it would be like building block— she would be starting at five instead of at zero. So at end of second day she would have ten points cumulative—five for day one added to the five for day two. End of third day, fifteen points cumulative.

"But if she don't come second day she lose three points of her five, and the next day two more. So the next Thursday she start again at zero. Except her body is still remembering to be stiff and sore from the Thursday before, so she got to work twice as hard just to get back to where she was when she started. Her body getting almost no benefit or change, and she getting no satisfaction. Soon she'll discourage and quit and call me thief for stealing her money.

"I am not trying to get rich by yelling at beginners to come every day. If they do or don't do is not going to make me rich. What makes me rich is seeing people lose pounds and years and get muscle tone and vitality and good health and bendable knees like Francis and a back that works like Archie and no rheumatism and easy childbirth and a brain that doesn't rattle.

"What I am telling you is, the only way you are really going to learn Yoga—which also happens to be the best and easiest—is to do it honestly and every day, when you beginner. Once you are very advanced, like

Leslie, then you can do it two, three times a week or even lay off for whole month and get your points right back 'cause you already got so many points cumulative stored in your body, like money in the bank.

"Until then, think of Cumulative and start building points, Terry Two. Don't try to figure it out, 'cause figuring is never gonna make you believe. Right now, like all beginners, you think I am the one with screws loose in brain. But if you do like I tell you, in a few weeks you will be giving the same lecture to your friends.

"Okay. Begin please. . . ."

14

1

Stand with feet together. Stretch arms over head sidewise like a Bird of Paradise raising its wings. Interlock fingers and make a nice tight grip. Release the index fingers. Straighten your arms completely, lock the elbows, press your arms against your ears, and stretch for the sky with everything you've got. You should look like New England church steeple. Keep your head up, your chin three inches away from chest min-i-mum for beginner, and your arms tight against the ears at all times. Inhale breathing for the first three parts; exhale breathing for the last part.

1

Unless you are fortunate enough to be made of cooked spaghetti, your attempt to raise your arms over your head and look like a church steeple will be as fraught with complications as inhaling and exhaling to a count of six.

The most frequent problem is that once you have managed to get your palms to-gether, your elbows straight, and your arms pressed to your ears, you find that your chin is crushed down onto your chest—in-stead of three inches away from it—that the muscles of your neck don't seem to be up to the task of lifting that chin, and that your steeple is a Tower of Pisa, pointing at the mirrors instead of the sky.

Attempts to right the structure produce various comic effects, the most common of which is that you somehow manage to move the shoulders backward and so point the arms almost at the ceiling, while the head refuses to budge, hanging forward ig-nominiously as though thrust through the stocks. Whatever the state your muscles are in, you will look and feel like one of Tennessee Williams's "no-neck monsters." But just keep forcing it. Progress is quick when you do.

2

Look at one point directly in front of you. Stretch your torso toward the ceiling as much as possible and, without bending your arms or legs, slowly bend directly to your right as much as possible. Keep your whole body facing front. If your upper body begins to twist toward the right, push your left shoulder backward and your right shoulder forward. Keep your arms straight, elbows locked, chin three inches away from your chest.

Simultaneously, push your hips directly to the left max-i-mum. Feel the beautiful pull along the left side of your body. Look in the mirror and see that your left side is making perfect, unbroken Half Moon. From the side your body is in a straight line. (If you were skinny you would disappear behind a beanpole.)

Stretch a little bit more, breathing 80–20, and stay there like a statue for ten honest seconds. The right side of your body has never been so happy!

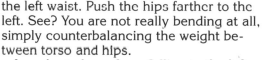

2

Feel like a no-**waist** monster? As solace, everyone, of whatever degree of fitness, has a great deal of trouble with this at first. What you are being asked to do appears foolishly simple—while involving virtually every joint, muscle and tendon of the arms, neck, torso, hips, and legs!

The difficulty of erecting the steeple and keeping it straight and perfect usually overshadows the rest of the Half Moon and Hands to Feet poses the first few days. You'll have a running start on the rest of it, though, if you recognize from the beginning that, appearances and spoken directions to the contrary, these are not essentially **bending** poses; they are **stretching** poses. (All of Yoga is essentially stretching, a fact which I will mention again and again.)

To help you get the feel of what I am saying, once your steeple is up and your chin is lifted as much as possible, sway your hips directly to your left **before** you do anything else. Sense the immediate pulling in

the left waist. Push the hips farther to the left. See? You are not really bending at all, simply counterbalancing the weight between torso and hips.

In order to keep from falling to the left as you do this, you must stretch the torso and arms harder and harder upward and to the right. The more the hips move left, the more your arms must stretch right and the more you appear to be bending. Yet you've actually moved only the hips.

If you can stretch only **one inch** to the side in this manner the first day you will be miles ahead of the fellow who breaks at the waist, twists his body, and bends halfway to the floor. He has accomplished nothing, while tomorrow, **you** will stretch a couple of more inches, and then more and more, and in a week you'll be doing a very presentable Half Moon and will probably have taken your belt in a notch or two.

3

Slowly come up to center position, keeping your steeple up like Atlas supporting world.

Once again lift the torso toward the ceiling as far as possible, keeping your arms absolutely straight and pressed against your ears, elbows locked, palms flat, thumbs locked, holding tight. Bend slowly to your left as much as possible, pushing your hips to the right max-i-mum. Do not twist body. Stay like a statue for ten honest counts.

3

This, of course, is an exact reversal of what you just did to the right. And here a curious fact will begin to reveal itself. You'll always be better at doing things on one side than on the other. But it will be a different side with each pose, lopsided creatures that we are!

Now a few words about "staying like a statue" for a count of ten. The exercises are called "poses" for good reason. The object is to reach the ultimate correct stretch that you, on that day, are capable of, then to hold that position for ten honest seconds. You then relax to allow the blood to circulate and normalize, then you stretch again. This alternation is where the benefit is derived. So don't advance and retreat like a nervous puppy. Freeze like a pointer.

Internally, however, strain ever more solidly into the "point" with each passing second. And always, on the count of ten, actively push it a speck farther.

4

Come slowly back to center position and straighten your steeple for the third part, the back bending. Lift your torso up out of your hips, inhale and hold the breath and slowly drop your head back as much as possible. Just let it go, slowly, straight back.

Slowly bend arms and body backward as much as possible, keeping arms straight. Breathe 80–20 now.

4

You'll find yourself tempted to do a lot of grunting and groaning at this point. Be my guest. Gasp, gurgle, hiss, croak, snort, or sob if you wish. Besides insuring the fact that you'll take a **breath** once in a while, noise will keep you company, increase your feeling of martyrdom and therefore self-righteousness, impress family and friends with your courageous efforts, and even impress the neighbors if you moan loud enough.

"Just let your head go" seems simple enough. Except that it won't seem to **go** back. This is due more to your own tenseness and timidity than to stiffened muscles. Surely you'll break, you fear, and if you let the head drop too far back, it will fall off because you forgot to screw it on.

Relax. Focus your attention on the base of the neck and let it go in that spot. More. You're still fighting it, all the surrounding muscles are tensed. The back bending

won't be half as difficult or uncomfortable if only you relax.

This portion of the Half Moon does, in fact, require bending along with the stretching or you'll never get there, but try to maintain the mental image of "stretch" over "bend."

5

Push your thighs, stomach, and hips forward as much as possible. You can bend your knees a little bit to help, I don't mind. Push forward more, bend back more. It is supposed to hurt a little in small of back, but it is a good pain, don't let it scare you. Your weight should be on your heels. Try to fall over backward but stop just before you do. Breathe 80–20 and stay like a statue for ten counts.

5

If you're like most people, you won't get too far the first couple of days. If you do bend your knees as I suggested, keep feet and knees together and pointed forward. Eventually **your** shoulder region will go that far back so that a platter could rest levelly on your chest.

To reach this ideal flexibility, a second point of relaxation must be recognized and utilized: When you have reached the stage where you simply can't seem to stretch or bend any farther back, zero in on the small of the back. Sense how the muscles there are tied in hysterical knots. Gather your courage, renew your effort to push hips and thighs forward, and just **let go** there.

The "exquisite" sensation you will receive will most likely elicit a cry that will bring the police running. But don't scare, nothing has broken, and you'll be doing it—burbling and gurgling with elation and anxious to demonstrate your new agility to any unfortunate soul you can trap for two minutes.

In any pose, the breakthrough comes when you learn not what to tighten, but what to **relax.** Yet that very relaxation is so difficult to achieve because of the very normal fear of hurting yourself. Realize, however, that not only is relaxation the "open sesame" to performing the postures, but it is also the way to be sure not to hurt yourself. (Witness babies, who can fall out of third story windows and just bounce, so relaxed are they, and drunks, who seldom seem to hurt themselves no matter what they run into.)

6

Slowly come up to center position, keeping your arms up over your head, elbows locked, palms together, and thumbs crossed.

We're now going into Pada-Hastasana, the Hands to Feet Pose.

6

You'll probably feel dizzy when you get there, and you will yearn to collapse your steeple and rest your hands on your head. But don't give in (and thank goodness the Statue of Liberty is made of sterner stuff than are we mortals).

IDEAL
Ardha-Chandrasana with Pada-Hastasana

7

Feet together nicely, lift torso upward, making sure your steeple is still pointing at the sky, not falling down on all the poor people. Now bend forward from the hip joints, all in one piece from tips of fingers to buttocks, legs straight, chin away from your chest, steeple perfect. Go as far down as you can this way.

When you can no longer keep the legs straight, completely relax, bend your knees, and reach around your legs, taking hold of your heels with your hands, thumbs and forefingers touching the floor. Bend your elbows and press the insides of your forearms completely against the back of your calves. Your goal is eventually to touch your elbows together behind your legs.

Now stretch your body as much as possible toward the floor, relaxing the coccyx. Touch your stomach on the thigh muscles, chest on the knees, and face below the knees—no daylight showing. Your goal is to

touch forehead to toes. Keep your eyes open.

REALITY
Half Moon Pose with Hands to Feet Pose

7

Feel the greatest stretch in the coccyx area at the base of your spine (you remember, that's where our ancestors had a tail). Increase the stretch by pushing your behind farther and farther to the rear while keeping your back and arms and legs as straight as possible.

As a beginner, let your knees bend as much as necessary to grasp your heels from behind, keeping your feet together of course. Your focus of attention is still on the coccyx. You relax downward from that point. That fact is important because the aim of this pose is to form a perfect standing jackknife. How does a jackknife work? On a swivel joint. In your body, that swivel joint is in the coccyx area, where you must eventually let go and relax.

There will be some of you who cannot reach your heels the first day, no matter how hard you try. Don't be discouraged, you are not alone. After a few days that won't be a problem.

As you lay your body against your legs, bend your knees even farther if you have to. If you can't get your body to touch your legs no matter how you slice it, then just do your darnedest to touch your forehead to your knees.

8

Now straighten your legs as much as possible. Straight! Lock the knees! Try harder! Concentrate on the exhalation as you breathe. Pull on your heels with all your strength and lift your hips toward the ceiling, forearms still pressed to your calves, hands holding your heels, head stretching for toes. From the side you should look like a Japanese ham sandwich, no daylight anywhere. I know it hurts behind the knees. Make it hurt even more. It is a beautiful hurt. So! Everyone looking spiritual and exhale breathing, hold pose for ten honest counts.

8

Unfortunately, nothing will accomplish the ultimate straightening of your legs except day-in, day-out determination and much strength. Try concentrating on lifting the hips rather than on the straightening and shake or shimmy the hips once you're down there grasping the heels. This shimmying loosens the muscles and lets you stretch farther and with more ease.

Because of the hurt you will feel in the backs of your legs as you try to straighten them, this seems the proper place to discuss pain. First of all, I am not talking about the pain of injury, disease, or sickness. I am talking about the step beyond discomfort—something you willfully put yourself through to obtain a happy result. (The first happy result is that the minute you stop the exercise, the pain stops.) This kind of pain is normal, expected, and good. It is merely your lazy, inflexible body protesting the sudden call to action. But if you have a special

medical problem, see the medical appendix for specific suggestions.

Yoga doesn't ask or recommend that you be a hero or a masochist. It asks only that you go as far as you can that day, try as honestly as possible at that moment. And so, as a general rule, stretch right up to any pain, then back off just before its threshold, and hold the pose there. Each succeeding day you'll have to chase the pain farther and farther to catch it.

As you become familiar with the poses and with your body's reactions to them, you'll learn what is merely laziness—just the body's needing to be coaxed, what to "push on through"—and what to halt at.

9

Release heels and come up slowly, exactly reversing the way you went down. Arrive at the starting center position with your steeple stretching beautifully to the sky once more. Now lower your arms slowly down to your sides like a graceful bird lowering its wings.

10

Rest a moment, arms easily at sides, legs and feet relaxed. Then repeat each of the parts of the Half Moon and the Hands to Feet poses, holding for ten seconds to each side, back and front. (This we call "second set.") Then rest again.

9

Please don't collapse when you get back to the center position. It's so undignified.

Additionally, the regimen of coming out of postures slowly and gracefully increases your stamina and discipline. Remember, I do not ask you to do anything without real purpose, or as they say, without method to my madness.

10

Cheer up. Besides having warmed your muscles, the first set will probably have exhausted you to a point of **relaxation.** Hence, you'll do the second set just a bit better, and as a consequence get a taste of Yoga's automatic feedback and instantaneous rewards. You'll really begin to understand about "instantaneous rewards" the second day, when you'll be able to do a dozen things you couldn't do the first. The third day the little triumphs will pile even higher. Yoga never fails to reward honest effort, and it does so with an immediacy that no other exercise, sport, or discipline can boast of.

And the day those legs finally straighten and there you stand having accomplished the "impossible," words can't describe the triumph and elation, self-satisfaction and sense of glowing accomplishment, self-confidence, and self-worth that will be yours.

Benefits

The Half Moon Pose gives quick energy and vitality; improves and strengthens every muscle in the central part of the body, especially in the abdomen; increases the flexibility of the spine; corrects bad posture; promotes proper kidney function; and helps to cure enlargement of the liver and spleen, dyspepsia, and constipation. It increases the flexibility and strength of the rectus abdominus, latissimus dorsi, oblique, deltoid, and trapezius muscles.

The Hands to Feet Pose increases the flexibility of the spine and the sciatic nerves and of most of the tendons and ligaments of the legs, and strengthens the biceps of thighs and calves. It also greatly improves blood circulation in the legs and to the brain, and strengthens the rectus abdominus, gluteus maximus, oblique, deltoid, and trapezius muscles.

Both poses firm and trim waistline, hips, abdomen, buttocks, and thighs.

Classnotes from Hilda

Hi. I'm the wizened little sprite you noticed when you walked in. I'm also seventy-five years old. And six months ago I wasn't any wizened little sprite, I was an arthritic old woman. But I didn't know how old till I came here.

The Half Moon we just did really opened my eyes. When I tried to raise my arms over my head and make that steeple, I could feel the old age, I knew where the arthritis was—and **why.** I felt it in the spine, the joints, and the bone structure, much more than in muscles or organs. That's the inflexibility that we let our bodies develop no matter how good a shape we think we are in, and no matter what our age. We all, even as kids, let our shoulders stoop. And the more years we let them stoop, the more they stoop, and the more the spine, those shoulders, and finally all the joints in the body get set "like in concrete," as Bikram says—one unbendable, rusted chunk from top to bottom.

Yoga, you see, is one big oil can. It's as simple as that. After six months of Yoga, I'm more limber than I was at twenty-five. Even my skin has firmed up and gotten moister. As for the arthritis, I scarcely feel it anymore if I do my Yoga a few times a week.

How old am I? I am **ageless.** That's what Yoga does for you.

The next pose, the Awkward Posture, works mainly on muscles and balance—but keep an eye out, too, for the oiling it gives the joints in your knees, ankles, and even your toes. You can't be **jointless** if you want to be **ageless.**

UTKATASANA
Awkward Pose

THREE

"Hey Bertha! Why you always late? Come on, hurry, we already finished Half Moon. Every day she leaves home five minutes earlier, least that's what she tells me, and still she always five minutes late. There is time warp between her house and here, you see. I think time warp is located in Bertha's brain.

"Know what was punishment for being late when I was a student? Punishment was, you had to sit on side and just watch class, you could not participate. I know, Terry Two, you are thinking that sitting on side would be nice thing to do right now. That's okay. You'll learn what is punishment and what is reward in this world. Nothing you get is good unless you do honest effort for it, and you cannot know the spiritual until you can control the physical.

"That's what is this Hatha Yoga I teach you. You learn to control the physical and be in beautiful health mentally and physically so you are not all the time running to doctors and whining to family, thinking small thoughts and considering your aches and pains and sniffles and screws loose in the brain. Hatha Yoga teaches you to tyrannize your body and make it your slave, so your body does not tyrannize you and keep you its slave. Only then, when you can control the physical, can you begin to know the God.

"There is no physical activity in world that exercises one hundred percent of your body, except this Yoga series I'm giving you. Jogging gets about ten percent, tennis fifteen percent, swimming fifteen percent, ballet thirty percent—none gets one hundred percent except this Yoga. When you come out from class, for first time ever in your life every organ, joint, muscle, tendon, nerve, and ligament—everything will be exercised and filled with fresh, oxygen-rich blood. Right now you're thinking you will drag yourself home. I make you money-back guarantee you will float home. You will think you are Mary Poppins and have to hang onto lampposts not to fly away.

"Come on, Bertha! Always I got to keep Awkward Pose waiting for you. Charlie, move over little, let Bertha in.

"This Yoga exercises one hundred percent of body, Terry Two, and you don't need swimming pool or tennis court or expensive equipment or years of training or even teacher in front of you. You don't need to be graceful or have talent or be athletic. Just you, a little space, and honesty is all.

"And it doesn't matter how 'good' you get at the positions. You don't have to become human pretzel—even though you'll surprise at how easy some of the pretzel positions are once you get limber. What matter is how good you get compared to how bad you were when you started. Comparing to yourself is single most important thing."

"Bikram?" It's a timid voice from a pretty but sadly overweight teenage blond.

"Yes, honey . . . Gail is it? This your third day, right? How you doing?"

"Okay—with the exercises, that is—but I've put on three pounds!"

"That's good. Means you're working hard at poses and so have big appetite. Is very normal to gain weight the first couple of weeks."

"But I want to lose weight, not put it on!"

"You will. Don't fright. Yoga restores all the systems of the body to natural, optimum working order. Very evenly, very gradually, your body will find the balance that nature wanted it to have, and fat is not natural. So your fat will melt away over the next couple of months. That's all there is to it."

Gail looks less than convinced.

"I know it sounds too good to be true, but you just talk to some of the other students after class. They will tell you. Forget diets, forget should I eat this or that and how many calories does it have. Just do honest round of Yoga every day for two months. Pretty soon, maybe third week, you gonna realize you don't think about food as often anymore, and when you do eat, you don't want as much as before, and the sort of food you eat will change 'cause your body is finding equilibrium, your glands and all your systems getting stabilized.

"But still you probably won't lose even a pound. You will notice instead that your clothes don't fit the same. 'Cause when you are doing Yoga your body is shuffling things around, taking some off here, adding it there. You can almost hear it humming to itself, it is so happy, like a sculptor with a new lump of clay. All of a sudden your friends will be congratulating you on the weight you lost when you have not lost one ounce—only inches. Then, when your body decides just exactly how it wants things to be, it will throw off the weight it doesn't have any use for, but so evenly and painlessly you will hardly know it. And when it gets to the weight that is exactly right for you, it will stop and maintain itself right there. Underweight is same, but in reverse. And with gland problems also. Proper balance and functioning of systems will restore proper weight."

"But some people have weight problems for neurotic reasons," says Florette.

"Didn't I just explain weight problems are because of improper balance and functioning of systems? I am not making jokes when I talk about fixing loose screws in brains. Your brain is nothing but a system in your body. As surely as to the spine or to the toes, Yoga brings balance to the brain. Also, Gail, doing Yoga gives you such feelings of self-esteem that you will want to lose weight and do it easily."

"That's right," Florette assures Gail. "You just get to feeling so gorgeous there's no way you'll let yourself stay fat."

"Enough. Begin please. . . ."

IDEAL
Utkatasana

1

Stand with your feet apart over six inches, heels and toes nice and square like an H. Raise your arms in front of you, parallel with floor, palms down, fingers together nicely, arms and hands with six inches between, muscles tight like rocks. Look at one point in front of your face and keep total concentration on that point. Keeping your heels flat on the floor and knees apart six inches, sit down until the backs of your thighs are parallel with the floor and stop there. (Pretend there is a chair behind you and you are sitting on it.) Exhale breathing.

REALITY
Awkward Pose

1

If you should find family or friends sniggering when you try this first particularly ungainly part of the pose, just invite them to try it. They'll stop laughing quickly enough, for it's as difficult as it is Awkward. Only the very limber can get thighs parallel with the floor on the first attempt, and then one must be careful not to allow the knees to creep together. Remember to keep them six inches apart. The hands also have a way of rising, so that they are no longer parallel with the floor, so keep a frequent check on them, too. The muscles of the arms and hands must stay tight. And please don't wiggle your fingers or toes!

Before Yoga

2

Now arch your spine back, striving eventually for a perfectly straight up and down spine position, as though your back were against a wall. To do this, put your weight on your heels, raising your toes off the floor and almost falling down backward as you arch your back. Keep toes, heels, knees, and hands all six inches apart. Stay there for ten counts.

2

I know it seems an impossibility at this point that you will ever be able to curve your spine and shoulders backward and sit straight up and down over your hips as though your back were against a wall, instead of thrusting your body forward as though about to dive into a pool for an Olympic relay. But you will, you will. And you'll never again have to wear socks to bed because of cold feet, since this pose sends fresh blood and oxygen to knees, ankles, and toes, which have been starved because no one's ever fed them.

After Yoga

3

Slowly come up. Still keeping hands, arms, and feet all over six inches apart, hands and arms parallel to the floor and tight like rocks, stand up on your toes max-i-mum, just like a ballerina.

Now bend knees and go down just halfway, forcing heels higher and more forward as you go. Make your spine absolutely straight as though against the Great Wall of China. Stop with backs of thighs parallel to floor, force heels even higher, and stay there ten honest counts, 80–20 breathing.

IDEAL
Utkatasana

REALITY
Awkward Pose

3

Those with flat feet will find this a task. But that's just the beginning. This second part of the Awkward Pose is a killer! However, for you ladies who want lovely legs, who have a few pounds of surplus cottage cheese hanging about on outer and inner thighs, or for you gentlemen whose legs have become sticks from sitting at a desk and who yearn for the fulsome muscles of your youth, this pose is it. Give it all you've got. Every time you think you've risen as high onto your toes as possible, lifted your insteps and heels and forced them forward as much as possible, you haven't. There's always a "higher" and a "forwarder." Work harder and harder each second until your legs are trembling with the strain. Eventually you'll rise nearly to a ballet point without the aid of toe shoes, and your legs will be the envy of every teenager on the block.

Also pay special attention to the Great Wall of China against which your back is to rest at all times during the pose. Don't lean forward. (But you will; we all do, even when we think we are perfectly straight.) Feel as though you are leaning just slightly backward. The farther forward and upward you roll the insteps and heels, the farther backward you must lean to be anywhere near straight-backed.

4

Slowly come up to standing position. Sink down on heels and relax feet but keep your arms up, straight, tight like rocks, and parallel to the floor.

4

The moment of rest will be an unparalleled relief for your quivering thighs. And those arms will be feeling leaden by now.

5

Next, rise up on your toes just a little bit, put knees together, and slowly go down all the way and sit on your heels, buttocks touching heels, knees still together. Keep your spine absolutely straight. Now push your stomach forward a little and drop your knees and arms down toward the floor a couple of inches so that they are absolutely parallel with the floor. From the side your body is in a perfect square. Stay there ten counts, 80–20 breathing.

6

Slowly come up just as you went down, knees still together, arms still up and parallel to the floor. Then lower the arms to your side and relax a moment.

Do second set, repeating each of the three parts of the pose and holding it for ten seconds.

5

Some of you might not be able to do a full knee bend at first, but **balance** is really the task of this one. The best of us still occasionally topple over backward. The tendency—once you've squatted down, gotten your buttocks in contact with your heels, lowered both your arms and your knees a couple of inches, and are leaning backward with your spine—is to let the arms and legs raise once more and to let the torso lean forward again. That is when you fall over. Concentrate on reinforcing the square, pressure always downward with knees and arms, always backward and upward with the spine, and you'll keep your balance.

6

The Awkward Pose gives some of Yoga's fastest results. It is positively inspiring to watch your legs change from day to day, to discover muscles, tendons, and definition that were never there before. I have seen people take twenty pounds off their thighs in a month with this; no exercise in the world is more effective for shaping legs.

Benefits

The Awkward Pose strengthens and firms all muscles of thighs, calves, and hips, and makes hip joints flexible. It also firms the upper arms. It increases blood circulation in the knees and ankle joints and relieves rheumatism, arthritis, and gout in the legs, and helps to cure slipped disc and lumbago in the lower spine.

Classnotes from Archie

It's true, this Awkward Pose does great things for the legs. Even at my age, on my sticks, I'm getting muscles again. But the important thing for me is that it helps things like slipped disc. Four months ago I was literally carried into this class on a stretcher. They had to lean me against a wall the first day to keep me perpendicular. And you think **you're** having a little bit of discomfort trying these exercises?

For nearly a month I couldn't move any part of myself more than an inch in any direction. But I reasoned I had nothing to lose, since the doctors had given me up. They didn't even think a spinal fusion would help. I'm still no great shakes at this stuff, but you could knock my doctor over with a feather. Compared to how I started, I'm a blooming contortionist! I still get twinges now and then, of course. And if I get lazy at the Yoga I really suffer. But, there's no comparison. I've got a life again.

GARURASANA
Eagle Pose

FOUR

"Lavinia, all the time you give up much too soon on all three parts of second set. On the second set you must always try harder than the first set. This is for two reasons. First, you are little bit tired from the first set, so it takes more effort to do the exercise again. But most important, the first set stretches you out, limbers you up, and gets the muscles warmed—and because you are tired you are more relaxed. All these things mean that you can go farther and do pose better the second time. To be lazy and not push farther is big waste of the first set, you see. Do you see, Lavinia?"

"Yes," says Lavinia with a perfectly blank face.

"Charlotte! How come you popping up from the knee bend like a jack-in-a-bag? I want always to see everyone coming out of poses as slowly as, and exact reversal of, how they went in. If you do not, you are not doing full pose and giving your body full benefit. Know what my instructor used to do if we rushed out of pose? We had to do it again, twice as long and twice as slow. I am very kind to all of you, but do not push your luck.

"Also, can be harmful to go into or come out of a pose too fast. Is single most sure way to hurt yourself in Yoga, to ambush your poor muscles, take them unaware. Doing a pose when you are not properly warmed up for it is another way to ambush

muscles. You know how blacksmith heats metal in fire before he hammers it into shape. Like blacksmith, is important to heat up muscle before you try to remold it. Just common sense, is no different from anything else. Football player, any athlete warms up. Juliet warms up slow and scientifically before she dances. Yoga is no different. So as a beginner, never make sudden movements. That way your muscles warm up naturally and are never subjected to a strain they cannot handle. If you do get to a place where they cannot handle it, you have opportunity, when you go slow, to know and to stop in time. If not, your poor muscles get frightened and tighten up quick, and you could strain. And that is nuisance, takes time to stop being sore when you strain or pull things, and so your progress is slowed.

"Yoga requires you pay exact attention to what you are doing. I am giving you exact scientific set of exercises and exact way to do them so you will never hurt yourself and so you will get optimum benefit—if you pay attention.

"Always, in Yoga, whether you are beginner or advanced, you are each day pushing your body just a little farther. You should think of it like you are an explorer going into new territory. Wise explorer goes slow, for he never knows what is around next bend. All our bodies, every one, is different from the other, no two in world is the same, and no one can predict where or when you going to find an Achilles heel, or knee, or hip joint, or vertebra. Only when you go slow can you feel it before you reach it, look it over, know how to treat it, how to make it stronger."

"Bikram," says Reggie, "if we're all warmed up and going away from any surprises, why do we have to come **out** of a pose slowly?"

"Because when you are holding your position for count of ten, your muscles, tendons—everything—is stretched and straining as hard as possible. To move fast with all your everything straining to utmost can pull something as fast as taking muscle unaware. Muscles and tendons are as vulnerable straining as they are when caught slack."

"I have a friend," says Bertha, "who pulled a muscle years ago doing Yoga, and now I can't convince her to come here with me. She says it's dangerous."

"Next time you see this friend of yours, Bertha, ask her did she ever know anybody that got hurt playing football, or tennis, or golf, or skiing, or ice skating, or skate boarding, or any other sport. Ask her she ever knew anybody got hurt dancing. Ask her she ever fell down while running or walking. Any physical activity in world, you can hurt yourself if you are careless or don't follow the rules. My goodness, more

people in this country destroyed their spines doing stupid things with hula hoops in one year than ever so much as pulled a muscle in eight thousand years of Yoga!"

"Have you ever known anyone to really hurt themselves doing Yoga?" asks Charlotte.

"Never with my method or method of my teacher. One million students at least been trained in this method. Not one person ever hurt—not when they followed directions. And anyone I ever knew, my students or my teacher's students, who get hurt even a little, who strain a muscle or pull something, it was because he did pose before warming up properly, or tried to show off his Yoga at parties, or scared muscles by coming out of pose too quick. Silly reasons like that. You got to treat your body with respect all the time. You warm it up nice, do things slow, and your body will always be nice to you.

"Important thing right now, Terry Two, is for you to understand that your body loves what you are doing. If it could, it would cover your toes with kisses of gratitude. Your body will not suddenly punish you with a torn tendon or slipped disc or some other terrible thing because you make it work and exercise a little. Only if you act silly and do Eagle Pose at party after glass of champagne and no warm-up does body get mad at you.

"So! No more Jack-in-the-bags. Begin please. . . ."

1

Stand with feet together. Look at your arms, which is right, which is left, don't mix them up. Bring your right arm under the left, crossing at the elbow. Twist your right hand toward your face and around the left forearm, tight like ropes, no daylight between arms. Place your right palm against the left palm, perfectly flat to each other, fingertips to fingertips. Force the palms counterclockwise so that both thumbnails squarely face your nose and the outside of your hands faces the mirror.

Keeping palms flat to each other and your chin up, lower your shoulders, and pull down on arms. Try to touch them to your chest and fit your mini-steeple nicely under your nose like an Eagle beak.

IDEAL
Garurasana

REALITY
Eagle Pose

1

As you might have noticed right off, this is not the most natural position in the world; every muscle and every bone is being asked to reverse itself and go someplace that, for a few days, it will flatly refuse to go. Men especially have problems because of larger biceps.

Try it this way: First extend your arms to the sides and then swing them together, wrapping the right arm as far around the left as the force of the swing will throw it. You will then find that the best way to work on the palm to palm is to get the pads of your fingertips together by whatever necessary contortion and, using your fingertips as leverage, push them against each other until—eventually—the palms will touch.

Making a proper Eagle's beak under your nose will be beyond you while you are still wrestling with the wrapping of the arms and the touching of the palms. But even if your hands and arms are not right yet, pull downward with the arms as hard as you can till the arms touch the chest. One of the greatest benefits of this posture is the flexibility and strength developed in the shoulders by this pulling. It also releases all those neck and shoulder tightnesses that cause stiff necks and tension headaches.

2

Keeping feet together, spine straight, and heels flat on the floor, bend knees about six inches until you feel a healthy pull. Look at one point in front of you and keep total concentration there so you don't fall over. Then transfer your weight to your left leg and slowly lift your right leg up high, bring it over the left thigh and wrap the calf and foot totally around like a rope, so that the top of your big toe hooks around the standing ankle just under the calf muscle.

2

The trick in getting your toe wrapped around under your calf muscle is an opposite technique to the "swinging" of the arms into place. Bend your legs and get yourself good and set. Then ever so slowly lift your right leg up really high and reach as far left with the whole leg as you can, lowering it on the far side of the left thigh and deliberately wrapping it as far around the left calf as it will go. Use the toes of your right foot, especially the big toe, just like fingers to reach and grasp with, even if you can't reach your ankle yet.

(Sarah needs to work a little harder on her flexibility—legs are not quite wrapped like ropes and hands should touch under nose.)

3

Once your right leg is wrapped as far as possible, sink down even deeper on the standing leg. Straighten your spine more, turning your hips to the right so that both hips are level and facing directly forward. Force your right knee right and your standing left knee left, so that you feel as though you could crack walnuts between ankles, knees, and thighs. Renew the downward pull of your arms, try to sink even deeper into the standing knee, concentrate on one point in front of you, and stay like a statue for ten seconds, normal breathing.

4

Uncross arms and legs and reverse posture to the left for ten seconds. Rest a moment, then do second set, to right and to left, ten seconds each, normal breathing.

IDEAL
Garurasana

REALITY
Eagle Pose

3

Some people have so much difficulty with one-legged balance at first that they have to steady themselves against a wall, get the legs wrapped as well as possible, find their balance, and then wrap the arms.

As a general rule, however, if you concentrate on one point in front of your face and then sit down as much as possible and straighten your spine, you'll find passable balance after a day or two.

Getting the hips square to the mirror and level, striving to sit ever more deeply, and creating the "nut-cracking" opposition in the legs are refinements that you will be able to concentrate upon fully only after you've found a solid balance.

4

Your initial success at this pose will depend upon the length of your leg from knee to ankle as well as your balance and flexibility. People whose forelegs are shorter must make up for lack of length by gaining even more flexibility than the rest of us. (But don't get smug, you colt-types; there will be poses where you will have to work harder because of your length.)

Benefits

The Eagle supplies fresh blood to the sexual organs and the kidneys, increasing sexual power and control. It helps firm calves, thighs, hips, abdomen, and upper arms. It also improves the flexibility of the hip, knee, and ankle joints and strengthens the latissimus dorsi, trapezius, and deltoid muscles.

Classnotes from Reggie

Here's a tip to fellow "desk jockeys." It's very easy to wrap your legs and get that toe around the ankle while you're at your desk, and no one knows that you're secretly practicing Yoga. This "desk practice" not only improves your flexibility and speeds your progress in the standing version of the exercise, but it keeps your lower half from going numb on the job as well. For that matter, the arm twisting isn't difficult to practice at odd moments either—and how it relieves neck and shoulder tension.

As a matter of fact, all of Yoga is a tension reliever, which is why I started coming every day. On top of that, Yoga gives me energy I haven't had since I was a kid. I need less sleep, my mind is clearer . . . Sure I "give up" close to two hours a day including traveling to and from class. But I feel so great that I can accomplish twice as much in a day now. You don't **lose** time by doing Yoga, you **gain** it.

A warning about the Eagle, however. No matter how good you get, **don't** use it as a parlor trick if you've been drinking. The bigger you are, the harder you'll fall, all wrapped up with nothing to stop yourself. Being inebriated and thus relaxed, you probably won't hurt yourself, but you'll break a lot of furniture, which can get expensive.

Also, I know you must be wondering, does the Eagle Pose really improve your sexual power and control? All I can say is, when Bikram promises you new life, he ain't just whistling Dixie.

DANDAYAMANA-JANUSHIRASANA
Standing Head to Knee Pose

FIVE

"Hey, Francis, that was great! Did everyone see Francis? He got his foot all the way around under the calf that time. **Perfect** Eagle Pose.

"How long you been doing Yoga now, Francis? Nine months? When Francis start, Terry Two, his knee was totally immobile, solid like rock. He had had an operation. His doctor told him he would never bend it again, that it was medically impossible to bend it. But luckily he trusted me when I told him I could make it flexible.

"I know about these things, Terry Two, 'cause when I was still in my teens hundreds of pounds fell on my knee and shattered it. As with Francis, the doctors were never sure whether I would walk normally. So I turned again to my Yoga, I had nothing to lose. In two months I was good as new.

"That's what Yoga is about, Terry Two. Yoga makes things that seem **im**possible, **poss**ible. You do magic without being magician. Right now you are watching your classmates and thinking it is impossible that you will ever do the poses like they do them. But you have only to listen to me and make pact to really try. Two months of Yoga every day and you will be as good as the best student here. What I mean by that is you will be in the most tiptop health **you** can be in. Anything chronic wrong with you—overweight, arthritis, bad back, old age, dandruff—Yoga will fix it. If you want to pray to your god so he help, too, that's okay. But better you save him for some really big problem. You don't need him for any of these little things, all you need is me and yourself and Yoga.

"Besides, there is a twelve-and-a-half-hour time difference, so he doesn't always hear.

"Do you know what Yoga means? Yoga is union of body and mind. It is discipline that brings together physical with mental, spiritual, and even sexual. Yoga is totally positive thing. With Yoga you add to yourself and thus to world. And so, when you do this Hatha Yoga, all the screws loose in your brain—which is always human being's worst chronic problem—get fixed, too. Because brain is just another physical system of body.

"You see, while you do Yoga is the one time, Terry Two, maybe first time in your life, when you absolutely and totally forget everything else: what you have to do at home, the pain in your toe, the fight you had with boss, the test you have to take, the bills, the anything. Whatever problem or pain you come in here with, in five minutes I give you so much else to think about you forget you even got it.

"You see how I never stop talking? That is so I keep your mind right here in this room. And I make you keep your eyes open all the time so your mind cannot wander off.

All you come in with—pain, anxiety, sadness, trouble—you leave it at the door. And for the time you do Yoga you escape into yourself where there is peace and rest.

"Right now you got perspiration rolling off you, your muscles are stretched farther than they thought they could, your shoulders are drooping 'cause you think you are tired—and you think I am crazy to sit here and tell you that you are enjoying peace and rest inside yourself. Answer me one thing, how many of your daily problems or anxieties have you thought about since we started our class?

"You don't have to answer. I know. And soon you will begin to understand why there are eight kinds of Yoga, and why I say you cannot begin to know the spiritual until you can control the physical. You will know why I say that these so-called Gurus, who start beginners right off in meditation, do not know what they are doing, or don't care. Minds cannot meditate when they are filled with bats and cobwebs and screws loose, and housed in junk bodies. Anyone who tries to meditate before all the body and mind are swept out and shining and oiled and no screws rattling, that person is just kidding himself and getting battier.

"It takes courage and intelligence, you know, to do the stages of Yoga right, and to start with this Hatha Yoga. Hatha Yoga takes more courage than any physical activity in the world. You are not kicking, throwing, or clubbing some little ball, or bird, or puck, or whatever. You are not locomoting headlong through water, or over ice, or turf, or a stage. You're not swinging, climbing, leaping, hopping, hanging, spinning, diving, or pedaling—or doing anything extraneous of yourself or with an outward direction. It's just you and nothing but you,

standing in one spot frozen like statue with no place to go for help or excuse or scape-goat except **inward.**

"Maybe it's a little bit frightening at first to have no help except what is inside you, Terry Two. But don't fright, don't scare. It is so peaceful in there, so good for your mind and your body to forget the outside world for this time you do Yoga.

"You hear my words now, but I know you don't understand them yet. That's okay. You got to **feel** the understanding happening in yourself, like a flower unfolding, to under-stand.

"The first thing you will probably notice is, you begin to look forward to each day's Yoga like it is cool oasis ending hard ride over the desert, not some terrible task. Your problems won't seem so bad, because you know you will be able to forget them for awhile during Yoga. And at end of Yoga you won't be half so angry or depressed or anx-ious or tired as when you started. Solutions to problems will suddenly be there, and you'll be seeing things in beautiful perspec-tive. You'll start smiling at people instead of biting their heads off, and new problems won't scare you. And you will begin to sleep good. Then pretty soon you'll start waking earlier—wide awake, your mind purring like happy cat, your body full of strength and energy. You will accomplish all sorts of things before the sun is even up and con-tinue through the day working more effi-ciently with body and mind than ever before.

"Then you will begin to understand what I mean when I tell you that this Hatha Yoga will tighten the screws loose in your brain and give you new life. It all happens inside of you, and you do it to yourself, Terry Two. You don't need me in front of you barking instructions. All you need are the words I give you here and your own honest effort.

"Okay. And so begin please. . . ."

1

Stand with feet together. Now lift your right knee up high and take hold of the right foot firmly with both hands, interlacing fingers firmly under your foot, about one inch from base of toes, with thumbs on top of toes. Don't let foot slip out of your hands throughout pose.

Straighten your standing leg absolutely and lock the knee. Keep your thigh muscles tight. Now pull the ball of your foot and your toes toward you with all your strength—this is a **pulling** exercise. 80–20 breathing when holding the leg.

Irene 1977

1

You may find that your foot is much farther away than you thought it was! But once you grab it, you will want a good grip, so if your feet or hands are wet with perspiration, wipe them with the towel first.

As a beginner, keeping that standing leg tight and locked is the most important thing, even if that is all you can do the first day. Also, the concept of "straightening" the leg is incomplete. You will find progress swifter if you think in terms of **bowing** the leg backward rather than straightening it. Look closely at the picture of the Ideal, and you'll see that the standing leg is most definitely curved backward.

Relaxation comes into play here. People tend to translate instructions to "straighten and stretch" as "tense and tighten." But not until you fully relax the standing knee, letting it bow right backward, will it really straighten. Don't fight it. Don't be scared. Let go.

Irene 2000

2

Straighten that leg out parallel to the floor, bending your elbows straight down toward the floor. Use your strength to pull more on the toes with your hands, push more forward with the heel, both knees are now locked.

Irene 1977

2

Don't be discouraged if you feel ten months pregnant and trying to tie your shoelaces as you attempt to straighten your raised leg out parallel to the floor. Even ballet dancers in tip-top condition have trouble doing it at first, plus not falling over in the process. So why should you, with your shriveled sciatic nerve and cast iron spine, be any different?

The fact is, straightening the extended leg has nothing to do with the leg itself; the leg is only there to connect your foot to your body. Your focus of attention must be on your foot, pulling hard on the toes until they point back toward you, pushing the heel forward with all your might. In addition, you will never get the toes pulled fully back, the heel thrust fully forward, and thus the leg fully straightened unless you—again—**relax**. Give it your honest effort each day. But be patient with yourself. This is a toughie.

Irene 2000

3

Now focus your gaze on one point on the floor, bend your elbows toward the floor, bend your torso forward, and touch your forehead on the knee. Exhale breath and stay there ten counts. (If you lose balance, try again immediately.)

Irene 1977

3

Do not attempt to bend forward and touch your head to your knee until you can straighten the extended leg and lock the knee. This is an absolute rule.

Do, however, get those elbows down toward the floor from the very beginning, almost hugging your leg, rather than letting them point outward like chicken wings. This makes balance easier, increases the pull on your toes, and will speed your progress in touching head to knee.

If you keep losing balance, it is because you are not keeping your gaze fixed, as though your eyeballs had turned to stone, on one spot either in front of you or on the floor. Experimentation will find the spot that works best for you.

To start getting your forehead down to the knee, use brute strength, gallons of perspiration, and whatever huffing and puffing makes you feel good. Once your muscles and tendons become fairly flexible and you are about halfway there, you can cheat a

little. Hold your position to the count of eight, bent as far as you can go. Then the last two seconds of the posture, pull your leg upward more strongly and reach with your forehead even more, trying to touch the knee if only for a split second.

You may fall over the first few times, but your body and your muscles will begin to remember and to figure out what they have to do to eventually make the contact and hold it.

Balance in the final stages of the Standing Head to Knee is accomplished by bowing everything farther and farther and pulling more firmly upward, toward, and, "into" yourself. Another trick to remember is relaxing everything in the hip joints, buttocks, and the lower spine.

4

Slowly straighten up and put your right foot down. Pick up your left foot, lock the standing leg, and reverse the pose, holding it for ten seconds. Then slowly straighten up and rest a moment before doing second set, to right and to left, ten seconds each side.

4

I know you won't believe me when I say that in a relatively short time, you will actually consider this one of the easiest poses. But mark my words.

Benefits

The Standing Head to Knee helps develop concentration, patience, and determination. Physically, it tightens abdominal and thigh muscles, improves flexibility of the sciatic nerves, and strengthens the tendons, biceps of the thigh muscles, and hamstrings in the legs, in addition to the deltoid, trapezius, latissimus dorsi, scapula, biceps, and triceps.

Classnotes from Bonnie

I don't think any of Bikram's students ever quite forget the moment of pure horror— the moment when first they get a look at the Standing Head to Knee. It is, visually, the most intimidating of the poses. Yet somehow, not too many weeks down the line, there you are doing it.

My best advice is, resign yourself and believe in "The Cumulative." Which is sort of like asking you to believe in Tinker Bell. But it works!

For over six months after I moved to Australia, I was "bedridden" with galloping laziness. I did no Yoga, just "lived on the interest" of my previous year of dedicated practice, till finally the neglect allowed my spine to kick up major trouble and I was forced back into it.

I found the first day I "did a class" that I still have Cumulative Points left. I could even do the Standing Head to Knee!

I won't pretend I wasn't sore for the next few days, but the point is that my body had forgotten none of the flexibility it had learned. . . . And now it was eager to forge ahead to new frontiers.

Yoga is the only discipline in the world that remains with you as an inherent part of your body structure once it is mastered. It is riches beyond rubies, which cannot be stolen away from you.

So dig in!

DANDAYAMANA-DHANURASANA
Standing Bow Pulling Pose

SIX

"How you feel now, Terry Two? Isn't that beautiful feeling in the back of your legs from the Standing Head to Knee? Don't laugh, I'm serious. When you were pulling on foot you were all uncomfortable and saying ouch and ugh and all words like that. But now that you have stopped, feel the blood coursing through your legs and how alive it all is. Your body is saying thank you. That is reason we all are willing to make it hurt for ten seconds, 'cause that little ten-second hurt gives our bodies hours and days of feeling wonderful.

"You Americans have a saying, 'Spirit is willing but the flesh is weak.' That is probably most untrue thing ever said. Some person like Lavinia invented that saying to excuse laziness. Each of us has been given a fantastic piece of machinery. That machinery can maintain itself in perfect condition for whole lifetime with nothing but proper lubrication. If you want to know how miraculous it is, think about how it struggles on for sixty, seventy, eighty years with no attention at all—abuse instead—cigarettes, alcohol, stupid foods, poisoning from air and pesticides, tensions like the god never intended it to have, and almost no exercise at all except getting in and out of car and pushing remote control button for TV. Sure, it gets sick and less efficient, it gets susceptibility to germs and viruses 'cause its natural defenses are weakened. It

gets weak spines and aching in the joints that have not been oiled in decades. And it begins to look and act in a way that we have given the name 'old.' But then you find out the most remarkable thing about it. With all that mistreatment, and no matter how broken down it is—I don't care if you are ninety-five or a Frankenstein monster from medical fusions—if you start oiling it properly again, if you do this Hatha Yoga, it will come back almost one hundred percent all by itself.

"Your body is not weak. Your body is strong! It is your mind that is weak, lazy, careless, selfish."

"Bikram," says a quiet, serious young girl, "is there a medical explanation for the energy you get from Yoga? You'd think it would tire you out—and when you're doing it sometimes you don't see how you can possibly get through another posture. Then when the class is over and you get up and walk away, you feel like . . ."

"Mary Poppins? Yes, Celeste, there is a scientific explanation, which also happens to be common sense. You heard me say often that in these Yoga exercises you use one hundred percent of the body, while things like tennis, jogging, swimming only get ten to twenty percent of the body. All sports, all exercises, even something like ballet—it is the same thing. These occupations do not exercise all the systems of the body because they were not designed, not intended, to. Whereas the Hatha Yoga was specially invented to lubricate, strengthen, and repair one hundred percent of your body machinery, including screws loose in your head.

"Just look what we already did in only five poses. First we started on the lungs, teaching them to breathe deep and be elastic as never before, and pumping lots of fresh oxygen into the circulatory system as well. Then we made you get the rust out of fingers, thumbs, wrists, forearms, elbows, biceps, shoulders, neck, and upper spine. To do this you had to use every muscle and tendon and ligament and joint and nerve and blood vessel that was there. Then we moved down the body, made you stretch the sides of your body from armpit to hips to thighs, and especially in the waist, and stretched your entire spine to right and to left. Then we moved around and, with the back bending, made you flex your spine backward and your pelvis and abdominal area forward. Then we leaned forward with the spine and moved down to the legs and tendons, at the same time compressing all the abdomen to accomplish other things in your insides.

"With the Awkward Pose we worked very much on your legs, the knees and ankles, and the feet and toes, while touching another area—your concentration, your balance—making your mind and nerves begin to exercise. The Eagle stretched stiff shoulders, hip, arm, and leg joints, and woke up sexual organs and kidneys. By then your whole body and nervous system were enough warmed up and ready for the Standing Head to Knee, which combines skills of the first five and works hard on your nervous system and mind, developing concentration, patience, determination, self-control.

"Also, in certain poses, you apply a tourniquet technique. That means you cut off blood to one part while forcing it very hard and in great quantity through another. Then you relax, allow all to flow normally again. This technique of hard exertion, then complete relaxation, is the key to the 'kingdom of health.' And the relaxation is very, very important. Do it as seriously as the poses.

"You see, Celeste and Terry Two, how scientific is this Hatha Yoga? From here, we will continue working systematically through your body leaving not one organ, bone, joint, muscle, ligament, tendon, blood vessel, nerve, or gland unoiled. The roots of your hair and your fingernails will feel it, your teeth, your eyes, your face. **All.**

"So when you work so hard in class sometimes you think you cannot go on, then find at the end you feel wonderful, this is because when you are all finished one hundred percent of your body is functioning at the optimum peak possible. Your body is singing with happiness. This happy feeling is called Energy.

"In minute I will tell you what then happens to your nutrition to give even more energy, but first we make your body write you even bigger thank-you note.

"Begin please. . . ."

1

Stand on your left leg. Turn right palm up toward the ceiling, and keeping it palm upward, reach around behind you. Bend your right knee and lift your right foot backward and upward, placing the foot in your cupped hand. Grasp the instep firmly about two inches below the toes. Your wrist will be inside the foot, your fingers will be pointing outward, and the sole of your right foot facing the ceiling.

2

Lock your standing knee, thigh muscles tight. Raise your left arm in front of you—elbow locked, fingers together and pointing forward. Keep arm close to your head throughout pose.

IDEAL
Dandayamana-Dhanurasana

1

Look carefully at the picture. Everyone gets this grip wrong at first.

2

As in Standing Head to Knee Pose, the standing leg **must** remain absolutely straight. So as a beginner, do only as much as you can while keeping your knee locked. Remember also not to let your arm drop. Think of it as a Siamese twin to your head.

REALITY
Standing Bow Pulling Pose

3

Now, level your hips squarely forward, raised knee pointing directly at floor. Focus on one spot in front of you and, in one solid piece from hip joints to fingertips, roll forward like a wheel until your abdomen is parallel to the floor. At the same time, point your toes and kick your right foot upward and backward with all your strength, straining against your cupped hand. (Don't let your foot slip out of your hand.)

Keep kicking up and back till you see your foot and leg sprouting right straight up out of the back of your head. Kick higher. Make it hurt a little in the back of your standing leg. Your goal is to kick up till you are in perfect standing split, both knees locked. (The higher you kick your leg, the more you will have to slide your hand down from your foot to the front of the leg.) Stay there like a statue for ten seconds, 80–20 breathing.

More

More

3

As impossible as this pose seems the first time you try it, it is the pose people seem to resent the least and are the most anxious to perform and perfect. It just looks so pretty, that even in the beginning stages you feel like Nureyev or Fonteyn.

The most important advice I can give you here is don't be in a hurry to dive into this position. So, once you have managed to stop hopscotching around on one leg, get yourself firmly set. Rivet your eyes on one spot, lock your standing knee, level your hips, and drop your raised knee toward the floor. Both of your thighs will then face directly forward, the bottom of your raised foot will point directly at the ceiling, and the toes will point directly at the back wall—all perfectly up and down, forward and back. No ballet "turnout" in Yoga. Only after you have done the above should you commence pivoting forward, at all times reinforcing the straight up-and-down, forward-and-back angles you began with.

Most important, don't let your lifted knee swing to the side—like a chicken wing.

Remember, the name of this pose is the Standing Bow Pulling; use your body exactly like a bow being strung and drawn by an archer. This means you must arch your head and spine ever more backward as you pivot forward. If you begin to lose your balance, raise your arm and head **higher** and kick harder upward and backward against your hand—in effect tautening the "bow" even more, or "picking yourself up by the bootstraps." You'll be amazed at how it restores balance.

Naturally, you must make it hurt in the back of your standing knee. But never **dive** forward or kick up exuberantly or abruptly. And never do the Standing Bow Pulling with cold, unprepared muscles. In other words, as nice as it is to have a spectacular party trick—don't.

IDEAL
Dandayamana-Dhanurasana

4

Slowly come back to center position. Reverse the pose on the left side, holding left foot in left hand and balancing on right leg. Hold ten seconds. Then return slowly to center position and relax a moment before doing a second set, ten seconds to left and to right.

REALITY
Standing Bow Pulling Pose

4

Once you get your abdomen truly parallel to the floor—and only then—will you achieve the graceful standing split shown in the photographs.

More

More

Benefits

The Standing Bow Pulling is a perfect example of the "tourniquet," or "damming" effect in Yoga, because it transfers the circulation from one side of the body to the other, and then equalizes it—circulating fresh blood to each internal organ and gland to keep them healthy.

Like the Standing Head to Knee, this pose helps develop concentration, patience, and determination. Physically, it firms the abdominal wall and upper thighs, and tightens upper arms, hips, and buttocks. It increases the size and elasticity of the rib cage and the lungs and improves the flexibility and strength of the lower spine and of most of the body's muscles.

Classnotes from Celeste

All the others have been telling you about the weight they lost, and how Yoga helps their backs, and so on. Well, I had other problems. Frankly, there were a lot of screws loose in my head. (If you're under eighteen, you're not supposed to have any serious problems, but when you are under eighteen, they sure seem serious enough!)

What Bikram was talking about earlier—that cool oasis that Yoga becomes, and the way you forget your problems while doing it, and how you're able to handle strain and tension so much better afterward and find solutions to problems—it **happens**.

He's a wise man, that Bikram. He never pushes the spiritual or meditative side of Yoga, only teaches the Hatha—because he knows that if you do the Hatha regularly, you'll realize the spiritual all by yourself, as naturally as the tail follows Mary's lamb.

That's one of the problems with Yoga, of course. It's the same as falling in love. If you haven't been there, you can only think you understand.

I have a theory about Yoga. Yoga is implosion. Theoretically, just as matter can be made to move outward with force and energy (**ex**plosion), so it can be made to move inward upon itself creating even **greater** force and energy (**im**plosion). Those mysteriously powerful black holes in the skies that astronomers are studying are the result of implosion.

Anyway, when, for instance, you follow Bikram's directions and pick yourself up by the bootstraps—move inward upon the Standing Bow Pulling to maintain balance—you are **imploding.** I'm not saying that, by moving inward into yourself, as you do in Yoga, you'll generate so much energy that you'll disappear in a big puff of silence, leaving nothing but a black hole in the middle of your living room. But you will discover huge, totally unsuspected sources of **strength,** of energy, of power.

Just **my** opinion, of course. But think about it as you do the next pose, the Balancing Stick.

TULADANDASANA
Balancing Stick Pose

SEVEN

"Susan, how come you didn't bring me chocolate chip cookies today, you mad at me? She make the best chocolate chip cookies, Terry Two. All chip, very little cookie—and chewy! I ask her for recipe. She won't give."

"If I gave you the recipe," says Susan, "you wouldn't need me anymore."

"'Course I need you, I always need you. What I do if I got no one to fall over backward in Awkward Pose to demonstrate to beginners how they shouldn't do it?

"Know why she fall over? All those chocolate chip cookies she didn't give me, she eat herself, and they go to wrong spot."

"Bikram," asks Archie, "how do you keep your shape with all the stuff you eat? Talk about a junk food junkie!"

"Don't you worry about my digestion. When you can do all that I can, then you can eat anything you want, too. I got perfect digestion, can eat tin cans if I want. Whatever I eat, I get almost one hundred percent nutrition out of it. Yoga puts me in such good shape I get all the nutrition or energy that there is to be had, burn it fully, and my perfect intestines show the rest of it to the door. Forty-five minutes, everything pass straight through."

"In that case," says Hilda, "you're five minutes overdue."

"Okay. Go ahead and laugh, but I am serious. What you got to realize is this. Re-member I told you Yoga improves the abdominal system, which includes your digestion. Most people—lazy, sedentary people, their systems all in disrepair, not operating efficiently at all—those people eat one pound of food, their digestion is able to turn only about twenty-five percent of that pound of food into nutrition. The rest is waste, but not all of it is eliminated. Much stays in the body as fat, or cholesterol, for instance, causing high blood pressure, heart trouble, constant exhaustion.

"As you progress in Yoga, your digestion gets healthier and more efficient. You are able to turn more and more of each pound of food you eat into nutrition, which means energy. You then need much less food than before, and less and less of that pound becomes waste that remains in the body to cause fat and disease and feelings of tiredness. That is why you also find yourself needing less sleep. Lots of sleep is a crutch your body need to restore energy only when your digestive system is not getting full nutrition from your food and creating energy. Hatha Yoga is the only physical activity in the world that increases energy rather than burning it up. Gets the energy even out of junk food.

"Nobody says that if you do Yoga you got to be some nut, eat only health foods, and not enjoy yourself, you know. In fact, it is the contrary. A well-functioning body is only too willing to turn any little calorie it can find into energy. Can you understand little bit now why it is the fat melts off you as you do Yoga?

"It is all so very simple, so common sense, there is nothing mysterious or mystical or hippy about what I am telling you. It is one plus one equals two. Without Yoga you got to eat three meals a day and sleep eight hours to keep your energy going. And what happens is that your energy gets less and less as the years go on because of all the waste, so you eat more and sleep more to make up. And your body gets even less efficient so you feel even more tired and get internal disease.

"Is exact opposite with Yoga. The more you keep at it, the healthier your systems become, the less you have to eat, the less you have to sleep. I sleep about two hours a night because I have one hundred percent energy from Yoga."

"Do you honestly only sleep two hours a night, Bikram?"

"Honest. All you businessmen like Reggie here, you all so concerned about efficiency and utilizing your time. Just think of all the wonderful things you could do if you did not have to spend half your life eating and sleeping—if twenty-two hours a day were at your disposal for working and playing and loving and enjoying."

"Please," says Archie, "we have enough of a population problem as it is. India, especially. Now we know why."

"You got one-track mind. You doing the Eagle Posture too often I think. I am sorry to disappoint you, but when you get so advanced in Hatha that you only want to sleep two hours, your mind is occupied with higher things."

The class hoots good-natured derision.

"Just for that, begin please. . . ."

1

Stand with your feet together. Stretch your arms up over your head sidewise, palms together nicely, elbows locked, and biceps touching ears, no daylight showing. Cross your thumbs and stretch your steeple toward the sky as much as possible.

Take a big step forward with your right leg, about twenty-four inches. Point your left toe behind you. Focus on one spot on the floor in front of you. Inhale breathing.

1

Get yourself set just as surely and solidly as you did with the Standing Bow Pulling. Straighten your steeple and move the head and shoulders back as far as they will go, farther than ever, allowing the chest to puff forward. That is exactly the torso position you are going to keep throughout the pose—at least you can **try**.

As in the Standing Bow Pulling, this is a totally front to back, up and down, parallel position. So, as you take your big step forward with the right foot, check your left hip. If you have allowed it to angle slightly left, adjust both hips and torso to face the mirror directly, and keep them level.

64

2

Now, keeping your hips level and muscles tight, in one solid piece from the fingers to your toes, pivot directly forward on your hip joints, raising your left leg behind you until your outstretched arms and torso and left leg with knee locked and toe pointed are completely parallel to the floor. Keep the knee of that standing leg absolutely locked. From the side you look like a "T." Breathe 80–20 and stay like a statue for ten honest counts.

IDEAL
Tuladandasana

2

As you pivot forward, use all your strength and determination. Be sure you are moving all in one piece (that means maintaining exactly the position you started in). To make this easier, pretend the floor is a pit of hungry crocodiles. Your standing leg is in no danger, being quite fortunately encased in crocodile repellant. But every other bit of you is in grave danger. The only way to keep your tummy and chest and left leg safe is to stretch your torso forward like crazy by lifting, ever lifting your arms and head, while you stretch more and more backward with the pointed foot and ever more forward with the fingertips, all the while lifting at front and back.

If you can pivot forward only two inches today, so be it. Tomorrow it will be ten. Remember to keep that left hip level. The more your press it down, the more pull you are going to feel in the back of your standing knee. This is good. Pull more.

Make up your mind you are going to do it for ten seconds, and don't give up! You at home, without me to growl and shout at you in person, are going to have to fight that temptation hard. Good luck.

REALITY
Balancing Stick Pose

3

Step back to center, feet together, arms still
stretched over your head, elbows locked,
biceps touching your ears, fingers interlocked, index fingers released together, and
thumbs crossed, nice tight grip.

3

Villain! You cry. Everybody wants to let their
hands collapse and rest on their heads for
an instant before doing the pose to the left.
For reasons of stamina and discipline, do
not succumb to this temptation.

4

Now reverse the pose on the left side, again pivoting forward on your hip joints all in one piece, from the tips of your fingers to your right toe now pointed behind you—until your whole body is parallel to the floor. Keep the knee of your standing leg absolutely locked. Hold for ten counts, 80–20 breathing.

5

Return to center, lower arms sidewise, rest a moment. Then do a second set, first on right leg and then on left, holding the position for ten seconds on each side. Then rest again.

4

By now I'm sure you realize how difficult it is not to feed the crocodiles.

5

Seldom will rest have been better earned.

Benefits

The Balancing Stick perfects control and balance by improving physical, psychological, and mental powers. In addition, it firms hips, buttocks, and upper thighs, as well as providing the same benefits for the legs as the Standing Head to Knee. It increases the circulation, strengthens the heart muscle, and stretches the capacity of the lungs. It is one of the best exercises for bad posture; strengthens the flexibility of latissimus, deltoid, and trapezius muscles; and improves the flexibility, strength, and muscle tone of shoulders, upper arms, spine, and hip joints.

Classnotes from Stephanie

Do you have any idea what it means to me even to be able to attempt an exercise like this—and to be able to do it almost one hundred percent correctly—after having pieces taken out of my spine and spending years in traction? Do you have any idea what it's like to **move** again, and do a thousand little things most people take for granted?

When you think of it, an hour or so each day really isn't so much to give to maintain a machine you expect to carry you around twenty-four hours a day for seventy or eighty years. I know people who spend that much time puttering with their cars!

What is it they say our bodies are worth? Nothing so grand as an automobile. Eighty-seven cents—probably a little more with inflation. Yet if something goes wrong with this machine worth under a dollar, not only can repairs cost tens of thousands of dollars, but there's no guarantee with the job—as I discovered when my operation plunged me into what seemed a lifelong purgatory.

I know how precious my eighty-seven-cent lump is, and I'm going to keep every penny shined!

DANDAYAMANA-BIBHAKTAPADA-PASCHIMOTTHANASANA
Standing Separate Leg Stretching Pose

EIGHT

"Do you know, every single one of you was holding your breath almost all the ten seconds that last set of Balancing Stick? Not one of you breathing 80–20 like I tell you to. What's the matter, don't any of you understand English? What words I got to use? How loud I got to shout? You are just making it harder for yourselves not to breathe. Not breathing contributes to not relaxing, for to hold breath you hold your muscles tight. This next posture I want to **hear** everyone breathe, with big emphasis on the exhalation. You will see then that on exhalation your muscles relax more and more, and you can sink deep and comfortable into the posture.

Is very important, breathing. There is whole part of Yoga called Pranayama, which is control of breathing. We start class with Standing Deep Breathing, which is part of Pranayama series. But for time being all I am asking is for you breathe **normal.**

"Another reason why you must breathe during the posture is to keep moving oxygen through the bloodstream so it reaches and nourishes the parts being worked on in each posture. Besides, no oxygen means muscles get tired quicker. More oxygen to muscles means more energy. To do the exercise while not pumping this good oxygen into the blood is to rob yourself of half the benefit of the pose."

"Bikram," says Charlie, "I just don't understand how you **can** breathe in the middle of some of these things."

"You do not understand how you can because you have not tried."

"I have, and it's hard! I certainly would not call it normal breathing, all twisted into a pretzel, and everything straining and unnatural."

"Why is it Americans insist in their minds on thinking of the poses as unnatural? There is nothing unnatural in this god's world. If it was unnatural it could not exist. What we do here are just things you are not **accustomed** to doing. And the surest, quickest way for you to make progress in Yoga is to make these poses your own—sit down inside them, have a nice cup of tea, look them over, and **breathe normal.** They are nothing to be afraid of. They are really your very best friends, eager to be comfortable, nice places to visit.

"All of you are counting the seconds while you are in the poses, with nothing in your minds but how fast you can get out of them. That is why you hold your breath. You have your eyes on the hole and never the doughnut. You can starve very quickly that way.

"You all do me favor, this next pose, do it like your body is a room you are entering. Look it over, be aware of everything. Keep your concentration on all the things that are there, what the pose is doing to each part,

70

how it is all feeling. Try to look in all the corners and under the rugs, open the closets, memorize the architecture, check the electricity and the plumbing, look for things that need repair or dusting or cleaning. Make it a leisurely visit, breathing deep and nice without fear, and be very surprised when I call you to come out.

"I am very serious. I want everyone to do this for me now. You are going into a room called Standing Separate Leg Stretching. Try to think of nothing but that room and give me detailed report of every little corner and closet that is in there when you come out.

"Okay. Begin please. . . ."

1

Start with feet together, arms at your sides. Take a big step to the right side, four feet min-i-mum. At the same time, raise your arms to the side, parallel to the floor. Point both feet forward, toes slightly turned in, legs straight, both knees locked. Exhale breathing.

1

The wider the stance you take, the easier the stretch. And the slightly pigeon-toed stance helps keep your feet from slipping outward and almost automatically makes your legs bow backward, putting weight on the **heels,** where it should be. You'll want to do this on a nonskid surface, though, to avoid any unplanned splits. Carpeting is usually best.

Those of you who have trouble taking a sideways step of at least four feet—keep trying, be patient.

2

Keeping legs straight and knees locked, bend forward from the lower back and slide your hands down the outside of the legs to the ankles. Grip the backs of your ankles firmly near the heel, all fingers together and thumbs on outside of feet.

Now, pulling on heels and legs straight, touch your forehead to the floor, as close to the body as possible. The weight is on your heels, very little on forehead. (Those of you who are touching your forehead comfortably, bring your feet a little closer together to make the stretching harder, and try to put your head right through your legs to the back of you.) Your goal is for your back to eventually be perfectly straight, not rounded. Stay ten seconds, eyes open, exhale breathing, and taking notes on the interior decoration.

2

Once you get the idea of this pose, your body position and weight will do most of the work, allowing you to just hang there and stretch. You could almost fall asleep. How to reach this mini-Nirvana? Good news. It's not hard. All you need is the patience to push it an inch farther each day, plus a few pointers.

The first day or two, once your legs are positioned properly, you might wish simply to put your hands on the floor in front of you, about twelve inches apart. Then, keeping your legs straight, bend your elbows toward the floor and roll your body forward like a wheel, reaching for the floor with your forehead. That way you begin to stretch yourself out and get the feel of the posture and the balance.

Now, it's obvious that to get your forehead to the floor you are going to need every inch of body you can find. Luckily, the body is hiding all kinds of extra stretching inches in tight muscles and tendons.

Once stretched out—do you remember Rubber Man in the comic strips? Like him, you will soon find yourself bending absolutely double from the base of the buttocks!

In my beginners' class, I teach only the methods outlined above. But since you are on your own, I offer you an alternate method.

Take your big step to the side, rest your hands comfortably on your thighs, and allow the pelvis, tailbone, and buttocks to relax. Feel that the weight of your torso is sinking right down into the crotch, forcing the relaxed buttocks ever more backward.

As the buttocks are forced backward, you will find the legs bowing more and more backward as well, and the torso automatically pivoting forward from the hip joints. As you pivot, let the buttocks continually relax further backward and lift your head and arch your spinal column an inch upward for each inch of descent.

Keep your spine straight as long as possible. Feel the hidden inches of potential stretch unfolding. As you travel downward,

IDEAL
Dandayamana-Bibhaktapada-Paschimotthanasana

3

Slowly straighten up. Right foot back to center. Rest a moment. Then repeat the pose for ten seconds.

REALITY
Standing Separate Leg Stretching Pose

slide your hands down the outsides of your legs toward the ankles. Once you can grab your ankles, use the strength of your arms to help pull yourself farther downward.

Finally, reach with your forehead and pull with your arms, trying to touch the floor by hook or by crook. You'll find that those last few inches to the floor will only be accomplished after you have relaxed the buttocks even more, and so shifted your weight fully back onto the heels.

When you've reached your limit, wherever it may be, hold it there for ten seconds, consciously relaxing the buttocks.

3

Once you are comfortably touching forehead to floor, shorten your side step to make it constantly more difficult so that you are stretching yourself more and more and getting your forehead touching closer and closer to the body, until it touches between the legs.

And didn't I tell you how relaxing it could be?

Benefits

The Standing Separate Leg Stretching cures and prevents sciatica by stretching and strengthening the sciatic nerves and the tendons of the legs. It helps the functioning of most of the internal abdominal organs, especially the small and large intestine, and improves the muscle tone and flexibility of thighs and calves and the flexibility of the pelvis, ankles, and hip joints, and of the last five vertebrae of the spine.

Classnotes from Charlotte

I don't think there is another posture in the whole series that so totally depends upon relaxation as the Standing Separate Leg Stretching. My awareness of that fact came about accidentally. Not that Bikram hadn't told me a hundred times. But in Yoga, you don't really understand a thing until you feel it happen inside of yourself.

For weeks in class with Bikram I had fought my way through this pose, using my hands on my ankles and my arm strength to pull myself downward, while concentrating on becoming a giraffe by stretching my neck to try to get the forehead a little closer to the floor, and clenching my teeth against the ache in the leg tendons. Then one day I got disgusted. I was exhausted from pulling, the tendons in my legs were on fire, and Bikram was making us stay down there longer than ever before, while he chatted with someone about oatmeal cookies. It just wasn't fair. So I stopped trying. I let everything go, relaxed, and hung there.

To my absolute shock, my leg tendons stopped hurting, the legs bowed back, something relaxed my lower spine—in fact, the whole area seemed to lengthen—and my forehead was suddenly only inches from the floor!

I stared in horror at the vast chasm of those remaining few inches.

"**Touch it!**" said the all-seeing Bikram. "**You Charlotte.** Touch . . . it!"

When Bikram uses that tone, it's an offer you can't refuse. I could never say afterward how I did it, but quite suddenly there I was—a tripod.

"That first time for Charlotte. Give her applause."

I came up wearing a silly grin, not at all embarrassed. I had earned the ovation, because a mild case of polio at the age of twelve had left me with a chronically bad back and very inflexible joints. Now there I was with my head on the floor.

And so my advice to you is, hang in there, Baby!

TRIKANASANA
Triangle Pose

NINE

"Marlon, you did that Standing Separate Leg Stretching just beautiful."

"Thanks," mutters Marlon, her tone indicating that she's afraid she knows what's coming.

"Just beautiful."

"Umhumm."

"But you know, all the others had their heads down between their legs looking at the back wall, so they couldn't see how the pose should be done in the absolute ultimate. Would you mind doing it once more to show them?"

Marlon shrugs, takes a large step to the side, puts her **hands behind her back,** palms flat to each other and facing upward as in a praying position, and then bends slowly forward, keeping her torso in an absolutely flat line from coccyx to the back of the head. Exactly like a jackknife, she bends double until her forehead is on the floor directly between her legs.

"Stay there, Marlon, it's good for you. Head more through legs. Sylvia, you do it, too."

Sylvia reproduces the demonstration, which puts a big hole in your cynicism. Sylvia could be anything from a dental hygienist to a lawyer, but she's certainly no dancer.

"You both breathing normal?"

There are two mumbled assents.

"You making notes on the interior decoration?

"Okay, you can come up. Charlie, what does the inside of your Standing Separate Leg Stretching look and feel like?"

"Pretty bleak. I used to think it was the sciatic nerves and leg tendons that were stopping me. But this time, really thinking about the insides, I could see the big problem is in the hips. I mean, I get to a certain point, then nothing happens. Zero. The strain runs up the legs beautifully, then those hips are like they were shot full of Novocain. I never realized it before."

"Okay. So that's where you got to concentrate. You have to mentally explore around your pelvic region till you find the right corridors, right doors, find all the light switches. What do you think you might stumble over while you're searching?"

"Well, the hip joints, of course, and the source of the sciatic nerves in the lower spine, a lot of organs in the abdomen, and the intestines . . ."

"Get to know those organs, make friends with them, especially your intestines. This pose is their very favorite one."

"Bikram." It's Celeste. "I've got a problem in my Separate Leg Stretching interior. It's been changing the last couple of days. It was all relaxed and comfortable, and I thought I had it licked. But now there's a pain in the back of my right knee—I mean a real pain."

"The sciatic?"

"That's sort of on the outside of the back of the knee?"

"Yes."

"This isn't it. This is on the inside."

"Oh. Okay, you got hold of the interior tendon of the leg then. This pose is the only exercise in the world that helps that tendon."

"Yes, but why is it suddenly hurting me?"

"You been having a lot of clicking in the joint of your right hip lately?"

"Yes. And it's kind of sore. I figured that was because it was getting more flexible."

"That's right. And you got to realize that 'getting flexible' literally means your bones and joints are changing. This is the thing that Yoga does for you. Does not matter if you are five or eighty-five, is only physical activity in world that changes your construction, from bone to skin, internally and externally, from the way you were born. You got to give your body time to accustom to these changes. If your hip joints are opening, it is going to make itself felt in the tendons, and maybe other places, too.

"Every single day, you will leave old aches behind and find new ones, stop snapping or clicking or popping in one spot and begin in another. Don't scare. It just means you found new frontier in your body and it is saying hello. Discouraging thing would be always to have same pains, same clicks. Would mean you were standing still, never progressing.

"Thing to do, Celeste, is to keep doing your Yoga every day. Worst thing for it when something is hurting is to lay off the Yoga 'cause you are scared, will take ten times as long to heal. Just do it very slowly, very carefully, and do not push—go only as far as you can without real hurt.

"Martha, you did it real good that time."

"I know! Keeping my mind inside, I really got the feeling of relaxation in the buttocks. I just looked it all over and told every single inch of me to let it all release."

"Feels funny the first time, huh?"

"Well—you're sort of glad there's no one sneaking up on you from behind. . . ."

"Not even old age can sneak up from behind now that you've learned to let your muscles feel vulnerable.

"This pose I do along with you, okay? Begin please. . . ."

1

(Do all that I say exactly when I say it, and not before or after in this pose.) Stand with feet together, arms up over head sidewise, palms together. Inhale breathing.

2

Step to your right with your right leg, about four feet, and at the same time lower your arms until they are parallel to the floor, palms facing downward.

IDEAL
Trikanasana

1

You're already becoming an expert at this.

2

You've been prepared by the Standing Separate Leg Stretching for the big sideward step, so be sure to take a huge one. If you don't, you will have to adjust your stance in the middle of the pose.

REALITY
Triangle Pose

3

Keep your left knee locked and turn your right foot and leg directly to the right. Push your hips and stomach forward and lean your upper body backward. Now bend the right knee directly to the right and—keeping your spine straight—go down slowly until the back of your right thigh is parallel to the floor. Your face, body, left foot, and hips are still facing front, and your hips are level. Keep the left leg straight and the left foot flat on the floor.

More

More

3

The injunction to keep your hips level and facing forward and the instruction to lunge down to a parallel position and/or hold it there will both seem ridiculous to you the first day. Cheer up. Things will get worse.

Before Yoga

4

Now, keeping your arms straight, bend your torso directly to the right, placing the right elbow in front of the right knee and the fingertips against the big toe, palm facing the mirror. Put no weight on your fingers. All fingers are together nicely and just touching the floor.

At the same time, look up toward the ceiling, twisting your head backward so that your chin touches your left shoulder, and point your left arm at the ceiling—elbow locked, fingers together, palm facing mirror. Stretch your left arm and shoulder higher. Your arms now form a straight line perpendicular from the floor to the ceiling. Viewed from the side, your entire body should be in one straight line. Looking serene and relaxed, like a flower opening to the sun, push stomach and hips forward as much as possible while twisting upper portion of body backward. Push the right knee backward with the right elbow. Stay there for ten very honest counts.

IDEAL
Trikanasana

4

Despite my directions, beginners usually do put weight on their fingers at first to keep from falling over or collapsing. Just try as best you can, though, to bear all the weight on the bent leg, which will probably be trembling from the strain. And if you feel like the Tin Man (or Woman) when you try to touch chin to shoulder, don't worry. Oil can is on the way.

Unfortunately, Terry Two, I cannot offer any shortcuts in this pose, only sympathy and understanding. The Triangle is quite simply a killer for most beginners. Just trying to hold it for ten seconds will at first preclude pushing stomach and right hip forward, and upper body and left hip backward, not to mention pushing the right knee backward. You'll be surprised after a week, though, at the strength you have developed. Then work on the refinements.

REALITY
Triangle Pose

After Yoga

5

Straighten right leg back to standing straddle position, turn right foot forward, keeping
your arms extended sidewise, palms to the
floor.

6

Now turn your left foot and leg to the left
keeping your right foot facing forward, and
repeat the pose to the left side for ten honest counts.

7

Straighten left leg to straddle position, turn
your body and foot to front. Move right leg
back to center, and rest a moment before
doing the second set to right and to left for
ten counts each. Then rest again.

Benefits

The Triangle is the only posture in the world
that improves every muscle, joint, tendon,
and internal organ in the body. At the same
time, it revitalizes nerves, veins, and tissues. It helps cure lumbago and rheumatism of the lower spine by flexing and
strengthening the last five vertebrae, and it
improves crooked spines. This is the most
important pose to increase the strength and
flexibility of the hip joint and of the muscles
of the side of the torso. It also firms upper
thighs and hips, slims the waistline, and improves the deltoid, trapezius, scapula, and
latissimus muscles.

5

This may require a derrick.

6

One side is always easier. If you're lucky,
this may be it.

7

Soon you'll begin to feel like that flower
opening to the sun. Just trust the process.

DANDAYAMANA-BIBHAKTAPADA-JANUSHIRASANA
Standing Separate Leg Head to Knee Pose

TEN

"Hey, Archie, why you were giving up before count of ten when we did Triangle to left?"

"Because you started talking to Sylvia on eight. I counted to twenty after that."

"Doesn't matter what you count, what matters is what I count. Ten does not exist until I say it. My teacher in India used to count like: 'seven, eight . . .' and then phone would ring. He'd go out of the room and pick it up and talk for five minutes, then come back in and say, 'nine, ten.' And if any of us had given up, we had to do over again. You guys got it easy, I am very kind to you. You even got nice soft carpet; we had to work on hard marble floor."

"When was that? Before the flood?"

"I wish. I wish I was that old. Some Yogis are hundreds of years old, you know."

"Oh come on, Bikram. Where? Produce one."

"What you think, I can pull one out of my pillows like rabbit out of hat? I will tell you the Ashrams in the Himalayas where they are. You buy your plane ticket, go there, and see them, talk to them."

"Can they prove they're that old?"

"Prove, prove, prove. Always people got to have proof. There are things for which proof can never be produced, but which your soul knows to be so. They don't need framed birth certificates. They have their birth certificates on their faces. Like when I was a boy, seven years old, I was in Benares in Northern India with my father and brothers—Asis, too—and all of a sudden people started saying a Yogi was coming to the bank of the Ganges, a famous man who was 280 years old. There were thousands waiting to see him by the time he got there. And he was ancient. The proof was with our own eyes. So what if his age was exaggerated, and he wasn't 280 but only 180 or even 120. It is still a magnificent thing. And there are many many like him, spiritual men in the Ashrams.

"But you don't need to go to Himalayas. Look at Hilda here."

"I'm not sure I like that!" says Hilda.

"Is a compliment. You look terrific. Like my teacher. At seventy he had figure like a boy, skin like a baby, energy that put us all to shame. Is no reason in world to grow old if you don't want to. You die when your Karma says so, but you don't have to grow physically or mentally old if you do Yoga. How are you feeling, Terry Two? Younger? Oh. Little tired? That's 'cause you've done less than half of the class. Poses are in scientific order to energize you fully, so after doing all poses you will feel younger, I promise. You might be little sore in morning, need derrick to lift you out of bed and lower you onto seat. You might moan and call me some names. But that is the very reason why you have to be sure to do class again tomorrow, no matter how stiff and aching your muscles are. If you do not do class you only get stiffer and hurt even more and lose benefit of today. But if you continue next day and day after, in four days soreness will go away.

"Bikram," says Sylvia, "I don't understand this business of stiffness. For instance, I wasn't at all stiff till after my third class. And now, even when I do Yoga every day, I'll suddenly wake up stiffer than I ever was in the beginning."

"Sure. Even I still get stiff sometimes. Look, there are no two bodies absolutely alike in this world, even identical twins. Each body is going to react different to same exercise. Also, our muscles and joints are all at different degrees of flexibility, and that can change from day to day in each of us, depending on if we catch a cold or if we've been under emotional strain or if we happy or sad or whatever.

"Usually, though, you get stiff when you use a muscle that has been too long lazy, or when you push a strong muscle a little farther than it has ever gone before. It is a very good sign to get stiff now and then after you have been doing Yoga awhile, means probably you are trying harder and going farther in the pose. Also, it makes you moan and groan in class, which is fun to listen to so I don't get bored up here.

"Okay. Begin please. . . ."

1

Start with your feet together, arms up over your head sidewise and make a steeple.

1

Don't worry too much about your steeple in this pose. The steps that follow will keep your arms straight.

2

With your right leg, take a big step sidewise to the right, three feet min-i-mum. Turn the right foot directly to the right, and this time you also turn hips, torso, face, and steeple directly right. Only your left foot is still pointed forward.

2

The wider the stance, the easier this pose will be, just as in the Standing Separate Leg Stretching. Also note the difference between this pose and the Triangle. Instead of keeping your hips and torso facing directly forward, you face directly to the side.

3

Keeping both legs absolutely straight, bend forward from the hips. Touch your chin to your chest and your forehead to your right knee. At the same time, touch the sides of your "praying" hands to your toes, with your fingertips touching the floor in front of your foot. Stretch your hands forward until your elbows are straight. Keep your eyes open, exhale breathing, and stay there ten counts.

3

You're having trouble? As a beginner, you are allowed to bend the right knee as much as necessary to touch your forehead to it. If, even bending the knee, you still cannot touch your forehead to it, you either suffer from a cast-iron spine or you are trying to touch your nose or chin or chest to your knee instead of your forehead.

This is a **forehead-to-knee** pose, and to get the forehead to the knee you must tuck the chin in and keep tucking it in while you curl toward the knee with the **forehead**, using everything you've got.

I don't know why, but no matter how many times I say the word forehead, my students continue to concentrate on stretching their backs, reaching for the feet, getting their chests close to their legs— everything but what I tell them. Always in these poses you must listen to me with your all three ears. Do as I say and you will not have to struggle half as hard.

All right. When finally you succeed in get-

ting your forehead to your knee and your hands down to your feet (which can take a day or weeks, depending upon the individual), begin then to use the forehead to actually push on the knee and straighten the leg back to the locked position. As you do this, you'll feel the stretch in the back of your right knee. This is **good.** Push more. Breathing will be a large help here. Big exhalations! And with each exhalation you will sink deeper into the pose.

Once you are all comfortable, forehead touching, both knees straight, work on twisting your hips even more to the right. Your goal is to face squarely to the right.

4

Straighten up, still facing right. Turn your torso, face, steeple, and then your right foot to the front but keep your feet apart.

5

Stretch your steeple and torso up to the ceiling as much as possible, then turn left foot directly left, hips, torso, face, steeple directly left, all but the right foot, which remains facing forward. Reverse the pose to the left, holding ten counts.

6

Return to center position as from the right, arms down sidewise, bring feet back together, and rest a moment before doing second set to right and to left, holding ten counts on each side.

4

Use this moment to relax your quivering right knee.

5

As with the Triangle, don't surprise to find one side much easier than the other.

6

This pose gives many of the same slimming and tightening benefits as the Hands to Feet Pose and the Triangle, so give it your best effort.

Benefits

The benefits of the Standing Separate Leg Head to Knee Pose are the same as those of the Hands to Feet Pose. It also slims abdomen, waistline, hips, buttocks, and upper thighs.

Classnotes from Quincy

Every pose you master in Yoga is a personal triumph that makes you want to stop people on the street and tell them the whole story. (Please don't, though. We've had two class members arrested for panhandling, and one young woman was hauled up on an even more embarrassing charge. It's the times I guess. I remember when Gene Kelly used to sing to people in the street, dance in mud puddles, stop traffic by tap dancing on roller skates—and no one minded at all. Now you can't even tell a fellow about your Yoga.)

Kidding aside, once your body recovers from its initial indignation, progress in this pose is swift.

Besides, the next position, the Tree Pose, is one of the least strenuous in the whole series. That gives you something to look forward to.

TADASANA
Tree Pose

ELEVEN

"Peggy, when your baby due?"

"Four months."

"Another of my students, Linda, she had her baby yesterday. She was in class in morning, and in afternoon she got her first pains, hardly had time to get to hospital before baby was born. You keep coming every day, Peggy. I make babies easier than any obstetrician."

"Would you care to rephrase that?" says Archie.

"You know what I mean. Peggy, you come every day, there is nothing so good for pregnant woman as Yoga. Like last two postures, and the next one we do, it is opening up your whole pelvis and hip joints, making all flexible, and making all your muscles and internal organs strong. Most women try to have babies with their stiff and inflexible bodies and weak muscles. They labor hours and exhaust themselves. But not you. Just at certain point in pregnancy you don't do exercises where you lie on stomach 'cause you will be so big and it would make too much pressure. But otherwise you can do everything else right up to last minute, like Linda did yesterday.

"Okay, begin please, Tadasana, Tree Pose. Feet together. Fix your gaze one point ahead of you, then slowly reach down and take hold of right foot, bringing it up in front of left thigh. . . . Lavinia!

"Okay, **everybody** hold it. Lavinia is not the only one. We have to get this straight. You, Florette, what was my direction after 'feet together'?"

"To pick up the right foot."

"No, it was not."

"You said to fix your gaze on one point ahead of you," answers Barbie.

"No one likes a smart kid," Florette warns.

"If you be as smart as that kid and just simply **listen** to what I say, and then **do** what I say, you could keep your balance and look like her, instead of hopscotching around the room like a child. Stand still! Why people don't listen? You cannot keep balance if you're wiggling around or if your eyes darting around like bird or lizard.

"I am going to show you how to 'fix your gaze,' Lavinia. For once in your life, you are going to listen and follow directions. And the rest of you watch. Everyone as guilty as Lavinia. You can do all the poses your **own** way, not following my directions, and you miss physical and mental benefits. No discipline, no concentration, screws still loose in brain, and muscles and joints still like cement. This time watch and listen carefully.

"Lavinia, just put your feet together, that is all, just feet together. Good. Now clear your mind, do not be afraid of the posture, do not think one thing. Relax your body, nice and easy. Now pick out a spot on the mirror that you like, something right ahead of you. Do you see something you like? Okay, what? What are you looking at?"

"There's a fingerprint on the mirror."

"I'll tell the cleaning crew to leave it there always, just for you. Keep on looking at the finger mark as though your eyes had turned to stone. Okay? Lavinia, now, and only now, you are finally ready to begin the pose. You are standing quiet and composed, your mind is at rest, and you have your complete and total concentration on your one spot. Is essential to use this technique before beginning any pose requiring balance.

"Keep your concentration, Lavinia. Shift weight to your left leg, and very slowly lift your right foot, and very slowly, easily, reach down and take hold of it with the hands. Slow. Eyes on the spot. Lift the foot up as high as you can, slow and steady."

Lavinia begins to wobble.

"Eyes on spot! It is your anchor.

"Foot slowly lift just a little higher if you can, then stay like statue. Nothing in the world exists but that finger mark on the mirror. **Don't give up!** Stay just a little longer.

"All right, Lavinia, you can put the foot down now. That was **perfect.** That was just beautiful. It's to see things like that, that I teach my classes. Oh, look at her big smile!

"And so, let us see if everyone can do it as good. Susan, you show them final pose. Okay, begin please. . . ."

Buddhadeb Choudhury in
Sankata Sana (Goodbye Pose)

1

Feet together, fix your gaze on one spot ahead of you. Balance on left leg and slowly raise right foot in front of left knee. Reach down and take hold of the right foot with your both hands, bringing it up slowly in front of your left thigh, the sole facing the ceiling. Lift the foot as high as you can, heel touching the hem of your costume close to the crotch, and rotate the sole toward the mirror, resting the top side of your foot high up on your thigh. Now straighten your spine, tighten your buttocks, and lock the standing knee.

2

Now force the bent knee downward. Your goal is to push your right knee down and backward until both knees are in one straight line. Keep buttocks tight and spine straight. Put your hands, palms together nicely, in front of your breastbone as if praying, and stay there ten counts, 80–20 breathing.

IDEAL
Tadasana

1

As you have just observed, balance can be your first problem here, but now you know how to solve it. You may also find it impossible, as yet, to get the heel up to the hem of your costume. Just do your best. Gradually, the flexibility in your hip and knee joints will increase.

2

The essential thing here is for that foot to be **high** and knee forced down as much as possible. And so, as pretty and dramatic as the praying position looks, as much as you yearn to look like my brother Buddha—ignore the temptation to pray and instead hold the foot securely in place with your left hand until the heel remains touching at least the bottom of your leotard/trunks when you let go of the foot.

Some of my students, after a year, still have to hold the foot up with one hand, putting the other hand up to the breastbone in half prayer. This is perfectly okay, as long as you keep trying honestly to keep the foot up there without either hand and are not just being lazy.

And if you have forgotten my instructions and have worn tights you'll make this pose harder on yourself. Your foot will slip on the material unless you already have terrific strength and flexibility.

REALITY
Tree Pose

3

Gently lower your right leg to the floor and relax it. Now fix your gaze, lock your knee, pick up the left foot, and do the pose to the left for ten seconds.

4

Relax knee carefully once more. Shake it out. Do not do a second set. Instead, immediately do the next pose, the Toe Stand.

Benefits

The Tree Pose improves posture and balance and increases the flexibility of the ankles, knees, and hip joints. By strengthening the internal oblique muscles, it prevents hernia. (This pose and the Toe Stand, which follows, are preparatory postures for the more advanced Locust Pose.)

But whether holding on to the foot or "praying," once you begin to get the knee pretty well forward and downward, you should start practicing the pose side-on to the mirror. In all probability, you'll be showing splendid progress in keeping the foot raised by allowing your buttocks to protrude.

So, **up** with the heel, **down** with the knee, **in** with the buttocks, bear the pain, keep your balance for ten seconds, and look ethereal.

3

Since your knee has been forced into unusual activity, treat it with care—shake it and wiggle it around a bit before reversing the pose.

4

The Tree Pose, more than any other, reveals the curious differences in human flexibility—and it often takes longer to perfect. Some people can't get their raised knee to move one inch in any direction. It seems at first that you are forcing the knee to work in a way that the designer never intended. Yet, when you begin to see what your body is capable of, and made stronger and healthier by, you'll begin to think that perhaps the designer did intend it, after all.

PADANGUSTASANA
Toe Stand Pose

Buddha Chaudhuri

TWELVE

"Lavinia, that was even better this last time.

"And Reggie, tomorrow you go out and buy pair of bikini bathing trunks for the class. I don't ever again want to see you in those Snoopy things. Know what he does? He cheat! I say the foot should touch the bottom of costume min-i-mum, and he puts foot up to droopy trunks and stands there grinning at me. Tomorrow, bikini."

"But Bikram, when I put the foot up that high and then push the knee down, the knee **hurts**!"

"'Course it hurts! It is not good thing for your body or your spirit when ripe, juicy apples fall off tree into lap. All good things in life, you should endure little mental or physical pain to earn. And strong knees is good thing.

"Think how you depend on your knees. Yet knees are weakest link in human being and hardest thing to strengthen. And these poor weak knees are never at rest. Think, for instance—joints of toes and ankles, hips, arms, fingers—all can rest in easy "natural" position, not do any work till called on. But what is "natural" position for knees? Knees are always being used for bending or supporting weight of whole body. So you must keep them strong and flexible, or in old age they will be one of your first big trouble spots."

"Bikram," says Celeste, "you said in class one day that the god is in the knees. What did you mean by that?"

"I said a thing like that? My goodness, the things I say to keep you lazy people moving. What I meant is, since knees are most difficult part of human to strengthen and control, person who has succeeded in doing that has developed five most important qualities of Hatha Yoga: faith, self-discipline, determination, concentration, and patience. These qualities you **must** develop to do these poses correctly. Then you are beginning to find the god, because the god is hiding in any little place that is difficult and painful, or weak and in need of control. Doesn't have to be physical. Can be thing like eating too much chocolate."

"I always did say chocolate is divine," says Florette.

"Oh my gosh. Come on, let's begin please. . . ."

1

Stand with feet together, focus one spot on the floor in front of you, and keep your total concentration there. Exactly as in Tree Pose, shift weight onto left leg and lift your right foot up onto your left thigh muscle. This time don't worry if it slips down a little. Lock your standing knee. Put your hands up in front of breastbone, palm to palm, praying position.

IDEAL
Padangustasana

Buddha Chaudhuri

1

It is handy that your hands should begin the Toe Stand in a praying position. Because if you've looked at what comes next, first you'll pray your knee will not break and then that you won't fall over onto your nose and become disfigured. Believe me, the praying is unnecessary. The Toe Stand is really the Lion Who Could Not Roar. It only looks fierce. (I still see doubt on your face.)

REALITY
Toe Stand Pose

2

Still focusing on one point on the floor in front of you, bend the left knee and lower yourself on the left leg as far as you can with your hands still praying. Then, bending from the lower spine, reach forward with both your hands to the floor. Supporting yourself with your fingers, sink down slowly the rest of the way on the ball of your left foot until your right buttock is sitting on your left heel.

More

More

2

You haven't really been asked to dive fifty feet into a half-filled tea cup, you know. **Don't** scare. Nothing will break. You have warmed up to this exercise, and gaining balance is probably your biggest problem here.

Once you have conquered your fear and become accustomed to sinking down using your hands for support, try to go all the way down without touching the hands to the floor. Your goal is to keep your hands in the praying position throughout.

3

Once down, with your leg still crossed over your left thigh, concentrate on one spot in front of you and move your hands to your sides, sitting on your heel and balancing yourself on the ball of the foot by using your fingers for support. When you begin to feel solid, put left hand up in front of your chest, using the other on the floor for balance. Then lift your right hand to your chest and stay there for ten honest counts, serene and praying for anything you want to, 80–20 breathing.

IDEAL
Padangustasana

3

The straighter the spine and the more parallel the crossed leg is with the floor, the better your balance will be. Also vital is concentration on that one spot in front of you. Balance in the Toe Stand is really only a matter of patience and concentration. Waving your arms as though you were directing traffic can be helpful in finding your balance. Also, learn to use the toes of your balancing foot just like fingers to grip the floor and help you keep your balance.

REALITY
Toe Stand Pose

4

Okay, stand up slowly by putting both hands on the floor in front of you. Push your left knee backward until it is straight and locked and come up the same way you went down. Lower right foot, shake the leg to relax it.

Raise the left foot onto your right thigh and repeat the Toe Stand on the left side for ten seconds.

5

Come up the same way again and gently shake out the left leg. Lie down on your back on your towel and relax for two minutes in Savasana, Dead Body Pose, head toward the mirror.

4

If the method just described hasn't worked for you after a few weeks, you can still practice the Toe Stand. Squat down, put one foot up on your thigh, and proceed to try for balance, first using both hands and then only one until you feel solidly set.

To bolster confidence in the strength of the knees, go back up to the standing position on one leg, foot still on thigh, by putting your hands on the floor, weight well forward, then rear your buttocks up and backward and push the standing leg back, locking the knee. Gradually, you gain confidence, balance, and strength.

5

Did I hear a sigh of relief when I told you to lie down and rest? Before you do so, please turn to the next chapter to find out how I want you to rest for maximum benefit.

Benefits

The Toe Stand develops psychological and mental powers—especially patience. Physically, it helps to cure gout and rheumatism of the knees, ankles, and feet. It also helps cure hemorrhoid problems.

Classnotes from Lavinia

I don't know if I can tell you how I feel at this moment, because there aren't **words** to the feeling. It's just . . . Mmmmmwaahh!

Do you remember the kid in gym class who was the only one they could never teach to do a somersault? The booby prize when they chose up sides? Well, that's always been me.

You'll never know what it took for me to come here the first day. I guess I just got myself into a trance, and all of a sudden here I stood in a leotard, facing Bikram. Within five minutes I loved that man with a passion.

Does that surprise you? Everyone thinks I hate him, everyone thinks that I refuse, with mulish stupidity, to come to class any day but Thursdays.

The truth is, I have **lived** for Thursdays. I wait like a child, hoping Bikram will shout at me. Because do you know what that says? **He believes I can do it.** He is the first person ever in my life to believe that I could do something graceful and athletic.

Why, then, didn't I come the other days? Because I didn't want to burst my bubble. If I came every day, soon, I was sure, it would be proved that I could not do Yoga. As long as I didn't allow that to happen, it couldn't ever be proved, you see. Bikram would keep believing, and thus my secretly graceful and athletic soul could still hope.

Now?

For ten seconds I balanced on one leg without falling over! I **can** do Yoga.

I **will** do Yoga.

SAVASANA
Dead Body Pose

THIRTEEN

"Nobody ever treats the Dead Body Pose like serious pose. Come on now, you all listen and give same concentration you give to Eagle or Standing Head to Knee. Relaxing is just as important part of doing Yoga as the poses that make you whimper and perspire. In each pose you stretch, squeeze, compress, twist, bend, release—which forces more blood into some portions of body and decreases and sometimes almost cuts off blood flow in other parts. When I tell you to come out of pose gently and rest a moment, this means muscles that were tensed relax, and blood flow gets equalized to all parts of body again.

"So it is very important after first half of class that you relax quietly on towel in Savasana, which is Dead Body Pose, for two minutes so blood, now loaded with fresh oxygen, can have a chance to feed all muscles and organs and tighten loose screws in brain.

"This is no joke, this tightening loose screws. Mind and body are two sides of same coin. Benefits flow equally to both sides. But if you don't relax, flow is stopped and both sides of coin miss benefits.

"So what I was talking about just before you did the Toe Stand, five keys to Hatha Yoga, are twice as important here. Because, funny thing, in Western society it is more difficult for people to relax than to work hard. People in your country feel guilty about relaxing. If they are not **doing** something all the time, they think they are bad people. Eventually, their minds don't know how to relax any better than their bodies.

"Relaxation is single most beneficial thing you can learn to do in this world. So apply even more faith, self-discipline, determination, concentration, and patience to the Dead Body Pose. You are fighting body tension plus guilt and impatience to get to next pose. You are fighting unruly mind that persists in flitting like butterfly. You must learn to make that butterfly sit perfectly quiet after taking nectar from its flower, doing nothing but being beautiful and digesting the nourishment that will make it even more beautiful and better butterfly. You must teach unruly butterfly mind that it is doing very difficult and good thing for itself. Tranquil inactivity of Savasana is full of more active and beneficial living than all the crazy rushing around you people do.

"And so, no more talk now. Begin please. . . ."

1

Spread your towel on the floor and lie flat on your back, feet toward the back of the room. (If you are doing the poses in front of a mirror, the mirror will always be "front of the room." If you are not using a mirror, choose one direction that will **always be** "front." Never should you lie with feet toward front of the room. This is for the rea-son, which you will soon understand, of developing a disciplined manner of performing my series of poses.) Arms at sides, palms facing upward, feet relaxed and flopping, no sexy pointed toes. Keep your eyes open and breathe normally. Breathing normally means forget about your breathing, how it is going in or out. Completely relax for two minutes.

IDEAL
Savasana

1

We never fully realize what reservoirs of tension we are until we are given the seemingly simple instruction to completely relax. Your hands will twitch and your feet are full of as much nervous energy as are your hands. And how about the muscles of your legs, buttocks, pelvis, and spine—especially your neck and shoulders—not to mention that convoluted gray mass called the brain? Suddenly you notice how many parts of you want to be tight because they're used to it.

The object of this pose is to consciously let go of as much tension as you possibly can. But trying to relax each part of you separately is akin to plugging a weakening dike. The minute you get your fingers pacified, the tension will pop up in your toes; if you manage to relax the buttocks, you'll find that your calf muscles have tightened, and so on. You could lie there and chase the tension for hours on end and still not catch and contain it.

It is far better to concentrate on relaxing the body as a whole unit. **Let the floor support you.** Pretend that all the spark has left your body. You would fall through space like a chunk of lead if the floor were not there, pressing upward, holding you easily. There's no need to worry, no need to be tense; the floor can do it all. **Let** it.

REALITY
Dead Body Pose

Benefits

The Dead Body Pose returns blood circulation to normal. It also teaches complete relaxation. This pose is done after each pose that follows.

Classnotes from Charlie

The thing that really galled me my first day in the class was that there I was, a big strong fellow, thinking I was in great shape—I mean, I've always played tennis, jogged, worked out at the gym, done a hundred push-ups every morning—and yet I was unable to do any of these poses so much as ten percent correctly. I couldn't even stand up! And I seemed to have the strength of a newborn pup.

Yet all around me were guys in nowhere the great condition I thought I was in who were doing the poses perfectly. And little fluffy females like Celeste and old women like Hilda were putting me to shame. That's a blow to a guy's ego that has probably ended many Yoga careers on the first lesson. Bikram said if I did Yoga every day it would take me two months to do the poses one hundred percent correctly. The way I felt, it would take two reincarnations!

I realized that there must be another kind of strength to be discovered in Yoga, perhaps a more important strength than the Charles Atlas kind. And what it boils down to, Bikram told me, is a balance between strength and flexibility. As Bikram says, "Never does nature make perfect person for Yoga or for any discipline. For every gift nature gives, there is problem to overcome. Who is strong is not supple, who is supple is not strong. Only discipline in world that can make you strong and supple both, and so give you **true** strength, is Yoga. It takes very hard work, but if you apply yourself, the balance you achieve between strong and flexible will guarantee health and vigor for rest of life."

When I heard him say that, I thought right away of a famous strong man—he could actually lift a car. But at forty his hip joints disintegrated. They just couldn't carry the weight of his musculature. Now if he had done Yoga along with his weight lifting and had made the joints of those hips as flexible as his muscles, he would truly have been a "strong man" and would be vigorous and healthy still.

Without that suppleness—despite your size and "strength"—you're really no more than a ninety-pound weakling.

PAVANAMUKTASANA
Wind Removing Pose

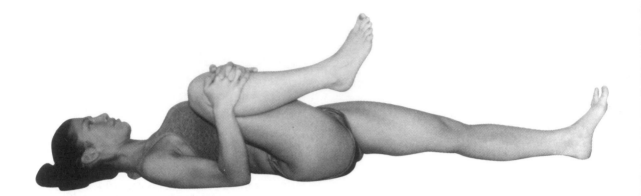

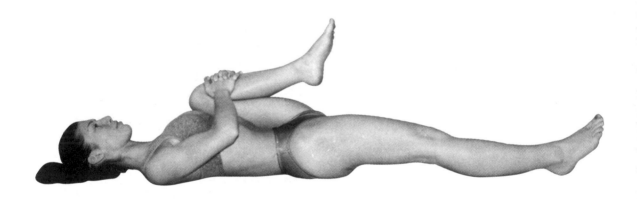

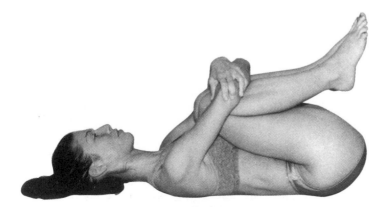

FOURTEEN

"How you feeling now you're all relaxed there in Dead Body Pose, Terry Two, good? I am glad to hear that, because all these things we have been doing till now was just the warming-up exercises. Now I start my class.

"Don't laugh. Why people always laugh when I say serious things? The more serious the things, the more they laugh.

"All the standing exercises, they are warm-up. The floor exercises are the serious Yoga. If you thought the stretching, squeezing, and twisting was a lot while standing up, you are in for big surprise at what we going to do now, Terry Two.

"Between standing exercises, we do not need to do formal relaxation. For the few seconds when you come back to center position between sets, shake yourself a little, and get ready for the next set, that is enough rest. But when you do floor exercises, is absolutely essential after each one to return circulation to normal and allow blood to flow evenly through whole body. Dead Body (I like to call 'Dead Man') is therefore one of most important postures, and we will do for twenty seconds between each of floor exercises.

"Remember I told you before about basis of Yoga—big effort, then total relaxation; big effort, then total relaxation? This is most true with floor exercises, and Dead Man Pose will teach you to relax."

"When I do my Yoga at home," says Bertha, "if I'm in a hurry, it's always the Dead Man I cheat on to get through fast."

"I know. And that is not good. When you do some of these floor poses, the pulse rate goes from 70 to 140 in five seconds. Even marathon runners, their hearts do not beat so fast after forty miles as after ten seconds of the Locust Pose. Is **good** for heart to speed up for short time. Is important way to strengthen the muscle. But after speeding up heart, you **must** do twenty seconds Dead Man so heart can calm down again, eat nourishing meal of fresh blood and oxygen.

"Did you have your blood pressure checked this week, Bertha?"

"Absolutely normal. The doctor took me off the medication. He still can't believe it's your system of Yoga that's done it. He says there must be some other variable."

"Tell your doctor send me a dozen of his high blood pressure patients. If they do exactly as I say every day for two weeks, I will show him what is variable."

"He's doing the next best thing. His nurse has high blood pressure, and she's starting class Saturday. She talked him into paying for her lessons as 'research.'"

"Smart pair. Doctor will see results.

"Okay, begin next pose please. . . ."

1

From where you lie in the Dead Man Pose, bend your right knee up toward your chest. Interlace the fingers of both hands and take hold of the raised leg two inches below the knee. Keeping elbows close to your body and shoulders relaxed on the floor, pull the knee down to your chest, foot relaxed, no sexy pointed toes.

Keep pulling the knee ever more firmly to your chest, until you feel it in your right hip joint. At the same time, lower your chin to the chest and keep it tucked in, head still flat on the floor.

The calf of the left leg stays touching the floor, foot relaxed. Every vertebra of your spine is now completely flat on the floor and there is pressure in the abdomen. Exhale breath and stay like a statue for twenty seconds.

IDEAL
Pavanamuktasana

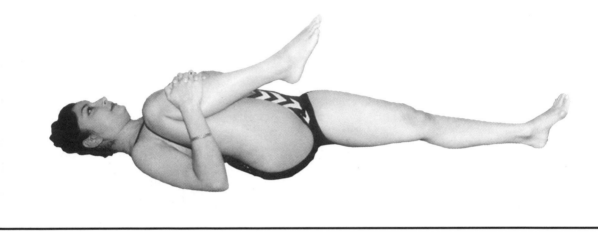

1

As a beginner, you'll find this easier to do if you move the leg a little bit outside of the body before pulling down to the chest. Don't be surprised, of course, if you can get the knee nowhere near the chest at first. Just pull as hard as you can, while concentrating upon relaxing, letting everything go, in the right hip joint. When you really try, there is fast progress in this pose. (If you don't feel the pull in your hip joint, you aren't really trying.)

It is essential to keep the calf of the left leg touching the floor. If this gives you difficulty, flex the toes up toward you; the calf will then touch.

Those of you who **can** get the leg down to the chest should use a more advanced grip. Instead of interlacing fingers, catch the raised right knee in the crook of the right arm, raise the left arm and grasp the opposite elbows, keeping them square, as though they were holding both knees. With shoulders on the floor, pull straight down toward your chest.

REALITY
Wind Removing Pose

2

Lower the right leg and both arms to the floor, then bend the left knee and reverse the pose to the left for twenty seconds, exhale breathing.

2

As in many of the other poses, you'll probably find more flexibility in the knee and hip joint on one side than on the other. So, keep urging the less flexible side by pulling harder, but with a slow, steady pressure.

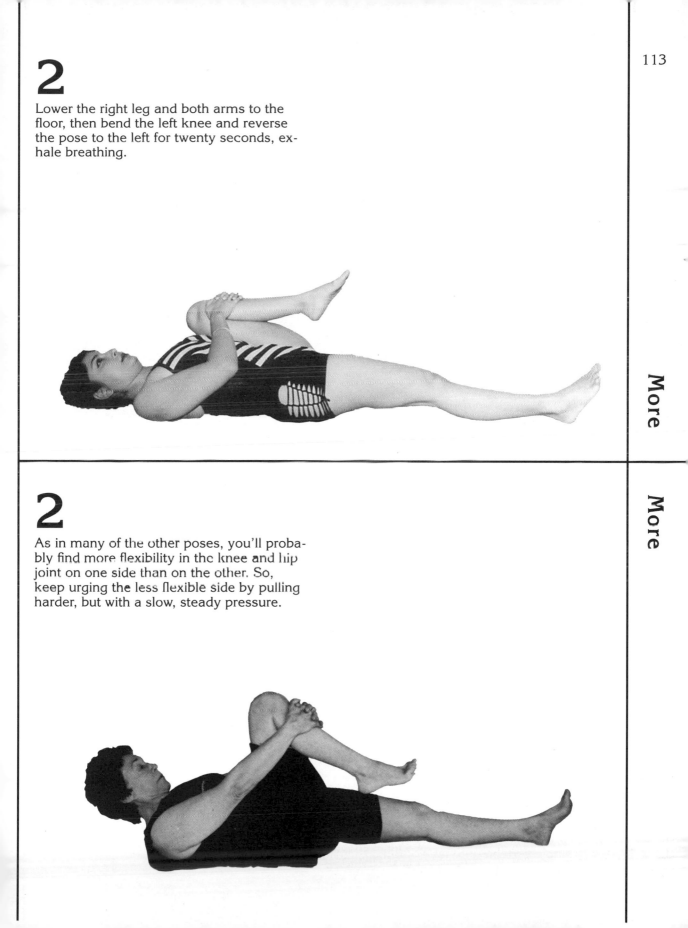

3

Now lower the left leg and both arms to the floor, then lift both knees up to the chest. Clasp your arms around them just under the knees and hug them tightly, hands grasping opposite elbows.

Keeping shoulders on the floor, pull your knees down to the chest as much as possible. Lower your chin to the chest, head on the floor, and force the hips downward until your tailbone touches the floor. Stay like a statue for twenty seconds, breathing normally.

IDEAL
Pavanamuktasana

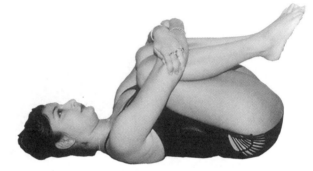

3

In this third part, if you are unable to get the legs far enough down toward the chest to grasp opposite elbows, then grasp forearms, wrists, fingers, a skyhook, or anything you can manage.

I'm sure you now see that all three sections of this pose are a bit like rubbing your stomach clockwise with one hand, patting your head with the other, while wiggling your ears. You have three separate and opposing things to think about—pulling down on the knees with all your might, keeping the chin tucked firmly down onto the chest, and either keeping the calf of the leg touching the floor or lowering the tailbone to the floor. While working on one task, you invariably forget the others.

A deceptive pose, indeed. It looks so simple, yet requires concentration and effort. Just keep in mind your two main goals: to open up your stiff hip joints and to push every single vertebra into the floor.

REALITY
Wind Removing Pose

4

Lower both legs and arms to the floor and do nice Dead Man Pose for twenty seconds, eyes open.

5

Repeat all three parts, holding for twenty seconds each, and then do Dead Man Pose again for twenty seconds.

4

As I said, you should feel a pull in your hip joints while you are doing the pose. But you may also feel the real effects when you release and lower your legs to the towel. So release slowly.

5

In the second set you are always more flexible, so remember the image of "letting go" in the hip joints and pull harder.

Benefits

The Wind Removing Pose cures and prevents flatulence, which is the source of most chronic abdominal discomforts. It also improves the flexibility of the hip joints and firms the abdomen, thighs, and hips.

Classnotes from Ralph

It's a crazy world we live in. The other night on a news broadcast I saw one of my colleagues, another M.D., spouting off against Yoga and warning people never to do it. I, on the other hand, urge my patients to run, not walk, to Bikram's class.

Bertha's doctor has just taken her off her last high blood pressure medicine, while other doctors swear you should never discontinue the medicine, no matter how long a patient's pressure is normal.

I've decided to install air conditioning in my house. I got five diametrically opposed plans for installation, all with different machinery and, of course, different prices.

Everybody's got their own ideas about things. And most try to shove them down **your** throat.

Not Bikram or his students. You want to believe that doctor on TV and never try Yoga? You want to try one class and never a second? Okay. We're sorry for you. Although Bikram will probably talk himself hoarse trying to help you, it's really all up to you. Once you try Bikram's Hatha Yoga, you'll either feel it or you won't.

SIT-UP

Maria Pogee

FIFTEEN

"I think I am going to give all of you a tip. I don't know why I should, but I am feeling very generous today. There is no move I tell you to do in this whole series of exercises that is chance, you know. Each thing has been exactly formulated for my beginners' classes and will make the Yoga simple and absolutely safe, if only you will pay attention and try to do exactly every little thing I tell you. I tell you listen all the time with your all three ears—I might as well hold my breath for all you listen to me. Try now to listen what I say.

"The next thing you going to do before the Cobra is sit up and touch your toes. I will ask you to do the same thing in a very exact way between a lot of floor poses. Treat the sit-up very seriously and try to go farther each time you do it. Because if you think about it, you will see that perfect sit-up is actually the Stretching Pose, next-to-last pose in class, that all of you moan and cry about and say is so hard. Silly thing is, many of you do perfect sit-up and still think you cannot do Stretching Pose. Explain that to me, please.

"So, throughout floor poses, everyone try doubly hard to do sit-ups and **think** about what you are doing. Then you will see, when we come to the Stretching Pose, if a miracle does not happen and you can do the Stretching Pose much better."

"Bikram," says Susan, "a friend told me that sit-ups are harmful for women. Is that true?"

"Oh my gosh. I think there are people who make career of sitting around and dreaming up ways to keep their friends from being healthy and happy. That is like the old idea that a woman should stay in bed for weeks after a baby, and not do exercise when she has her period, and ride sidesaddle, and other silly old junk. The only way you could hurt yourself even little bit in any of these exercises is by not doing it **exactly like I tell you,** under the conditions I tell you.

"As a matter of fact, in my opinion females even more than men should do sit-ups every day, from childhood even, just especially to keep all the abdominal muscles and female organs strong. Who was this friend told you this silly thing?"

Susan blushes. "My mother."

"Oh. Well, a mother is a good person to keep for a friend. You bring her and her fears to class next time and we will get her doing Yoga and change her mind about sit-ups.

"Okay, everybody try a practice sit-up; do exactly everything what I say. . . ."

1

From where you are in Dead Man Pose, raise your arms up over your head simultaneously inhaling, and sit up, keeping legs straight and heels on the floor. Use the force of throwing your arms toward your toes to help you sit up.

Just before you reach upright position, start exhalation and dive forward, reaching for your toes, which are flexed back toward you. Grasp them, laying your whole body and face out flat on your legs or at least touching your forehead to your knees. Touch the floor on either side of your legs with your elbows. Exhale breathing.

Maria Pogee

IDEAL
Sit-Up

1

Some people at first can find no way in the world, no matter how hard they try, to sit up—as though something big and fat is sitting on their chest. Others can do it, but their feet bounce two feet into the air as they do so. (This is all right. You can even lift the feet farther and use their downward thrust to sit you up if you are having a great deal of trouble.) Others can sit up keeping the feet firmly on the ground, then can't grasp their toes, much less touch even the forehead to the knees.

Do not let your particular state of unfitness discourage you. Give each sit-up your honest effort and in two months maximum you will do it exactly as I described.

REALITY
Sit-Up

Benefits

Sit-ups strengthen and tighten the abdomen and increase the flexibility of the spine.

BHUJANGASANA
Cobra Pose

SIXTEEN

"Valerie, tell Terry Two about your neck."

"Well, I came to Bikram's class six months ago because I thought Yoga was religious. I thought I was going to meditate."

"I know. That is the reputation, the label people put on Yoga. People don't realize there are eight different levels and sixteen stages of Yoga. Lots of variety like Howard Johnson's flavors."

"I was worried when I found out it was exercises," says Valerie, "because I had just gotten a whiplash in a car accident and my doctor had told me never to bend my head backward, never to lift anything heavy, never to bend forward—or do anything! I'd spent three months and hundreds of dollars in physiotherapy and my neck still hurt. But I decided to trust Bikram. And it was amazing. One solid week of this class did more for my neck than a whole summer of traction, whirlpools, and massage! And I haven't so much as had a pain in my neck now in six months of Yoga."

"Of course. Proper kind of careful stretching was what you needed. But Western medicine is funny. Is slow and stubborn to accept Eastern cures we have known for thousands of years. But little by little, doctors are beginning to pay attention to Yoga. Lots of doctors came to my classes to investigate. Many now my students and even send their patients here.

"Like Ralph. What you a doctor of, Ralph?"

Ralph grins. "I'm a rare bird. A GP."

"You're like me. You know importance of treating whole body.

"Okay, time to do Cobra. Do beautiful sit-up, turn around on your towel to face mirror, and begin please. . . ."

1

Lie down on your towel on your stomach, legs and feet together, all muscles of legs and buttocks tight like rocks, toes pointed. Put your palms flat on the floor directly underneath your shoulders, fingers together and pointing forward, fingertips projecting no farther than the tops of the shoulders. Keep your shoulders down naturally and bring your elbows in until they touch your sides, and make sure they touch sides throughout pose. Inhale breathing.

IDEAL
Bhujangasana

1

As soon as you tighten your muscles, you may be attacked by cramps in legs and feet, the cramps continuing throughout the floor poses. If it should happen to you, grin and bear it. Flex and wiggle the affected parts, then renew your efforts. The cramps subside as the days go by.

REALITY
Cobra Pose

2

Now look up at the ceiling. Using your spine strength to lift yourself—rather than pushing with your arms—raise your torso off the floor just to the belly button.

2

You probably won't be able to find a spinal muscle in that disused maze back there, much less mobilize one. As a hint, the muscles you use to arch backward when you have a backache are the very ones that lift the torso in the Cobra.

Now, just for the fun of it, try lifting the torso without putting any weight at all on your hands. You might even raise the palms slightly to prevent cheating. This will enable you to feel the muscles of the lower back and understand not only how weak or strong they are but recognize the "contact" that must be made to eventually go all the way up with spine strength alone.

Whether you can lift five inches or no inches without weight on the hands, you do have your hands and arms to fall back upon. As a beginner you will make good use of them.

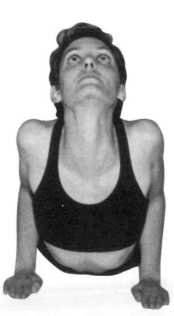

3

Now arch back as much as possible with head and torso, still using the strength of your spine. At the same time, press the belly button into the floor. Your elbows are still tight to your sides, your shoulders are relaxed and down.

When viewed from the side, the angle inside your elbows should be ninety degrees. Breathing normally, stay twenty seconds, your face looking serene and content, 80–20 breathing.

IDEAL
Bhujangasana

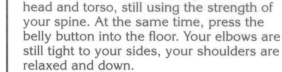

3

Maud is doing a pretty Cobra, but her elbows are only seventy-five degrees, not ninety. She can do better.

In this step, the essential point is to push that belly button through the floor with everything you've got, while arching spine, neck, head backward and releasing the small of the back. And glory in the stretch. Feel what it is doing for your waistline. Abandon worries about low-back pain and double chins. Your friend the Cobra has come to save you.

REALITY
Cobra Pose

4

Slowly lower the torso, turn your face to one side, and relax on your stomach for twenty seconds. Keep your eyes open, your arms down at your sides, palms upward, and heels relaxed outward.

5

Repeat the posture for twenty seconds and then relax again for twenty seconds.

4

Do not collapse back onto the towel. Use your spine strength and arms to lower yourself smoothly.

5

This is not a hard pose requiring great strength or unusual contortions. You're not going to hurt yourself, you're not going to strain anything, there is nothing to fear. What the Cobra takes is willpower—a commodity that is usually in shorter supply than strength. Barring medical problems, slow progress in the Cobra means just one thing. L-A-Z-Y!

Benefits

The Cobra is one of the best ways to maintain the body in perfect condition. It increases spinal strength and flexibility, helps prevent lower backache, and helps cure lumbago, rheumatism, and arthritis of the spine. It also relieves menstrual problems (irregularity, cramps, backache), cures loss of appetite, helps correct bad posture, and improves the functioning of the liver and spleen. The Cobra strengthens the deltoids, trapezius, and triceps.

Classnotes from Peggy

I'm going to be really unhappy when I get too big to do this one. It's supersensational for the spine, getting rid of rheumatism and arthritis—and when you're pregnant, boy oh boy, can your spine feel arthritic!

It's also great for menstrual problems—irregularity, cramps, stomach and backache—many of which are the same discomforts you just live with when you're pregnant.

Aside from the relief from discomfort that Bikram's series gives, the strengthening of the muscles and internal organs and the opening and flexing of the pelvic and hip joints are really a godsend in pregnancy. His method also teaches you mental and physical stamina.

And every woman I've talked to who has gone through her pregnancy with Bikram says the same thing: The delivery is almost effortless and they feel great. In fact, they get right back to class and have their figures back in a couple of weeks. The babies even seem exceptionally healthy, probably from all the super nutrition that was in the mothers' blood from Yoga.

Honestly, what more could an expectant mother ask for?

SALABHASANA
Locust Pose

SEVENTEEN

"I see some of you are closing your eyes as you lie there. For time number fifty-two-thousand, I tell you **eyes open.** It is essential to keep your concentration here in this room and in your body. If you close your eyes your mind goes wandering off. You think about your hair is gonna be a mess after class, the phone call you forgot to make, the supper you gotta cook, the bill you have to pay—all these things that seem so important but are really of no importance whatsoever. In a week, will it matter what you have for supper tonight or if your hair got messed? In ten years, will you remember the phone call you forgot to make? In thirty years, what will be importance of the bill you have to pay?

"What **will** matter is the body that you expect to carry you through those years and the health you do or do not have.

"And very much it will matter what the mind inside the body is doing, how it is thinking, whether you are miserable and tense and unhappy or at peace with yourself and with the world and so beautifully happy.

"This is what you have a chance to do with Yoga. Only follow my words, keep your eyes open, keep concentration in your body—and you will come naturally to peace in your mind."

"Bikram," says Barbie, "where is the third ear we are supposed to listen with?"

"It is in the very center of the everything you are, Barbie. It is every little part of you, and the whole of you all at once."

"Sometimes I think I can hear with it, and other times nothing happens."

"To hear with it really, truly, you must be completely empty. That is why I keep problems and thoughts of outside world away from you here in class, to allow you to be empty, like glass into which the god is waiting to pour sparkling pure water of the truth. If you are three-quarters full, the god might be able to pour in little bit. This is better than nothing. But the gift gets much diluted with what is already in the glass. You got to have your glass empty and clean to receive pure truth, real hearing with the third ear.

"So each day here in class we work on that, on emptying the glass just a little more, so you can hear better and better. When finally you are completely empty, and we wash you in Lemon Fresh Joy so you can see your own reflection, then you will fill with truth and you will hear with clarity and understanding, hear beautiful things you did not think possible.

"Okay, begin please. . . ."

1

Lie on your towel on your stomach, with your chin on the towel. Put your arms under your body with elbows turned upward against abdomen and palms flat on the floor, your little fingers touching. (Look at the photograph, which shows the proper position of arms and torso that floor would see.)

IDEAL
Salabhasana

1

The undulations, heavings, rollings, and gruntings as you try to maneuver your arms into this new and unusual position are marvelous to see and hear. Try this technique: From your position of relaxation on your stomach, push your right toe into the floor to lift right hip and roll your body slightly to the left. Then slide the right arm directly under you with the palm flat on the floor. Lower your right hip onto the hand and arm. Then push with the left toe, roll your body onto the right arm, lift the left hip, and slide the left arm under you, palm down. Get the little fingers touching and elbows as close to each other as possible. Then lower your left hip onto the left arm.

You will thus have both arms nicely pinned, and you will suddenly feel like a trussed goose. Your head will be bobbing around trying to look casual, and your elbows will most likely begin to protest the position in which they find themselves. Put your chin solidly on the floor and wait for what follows.

REALITY
Locust Pose

2

Chin on the floor and feet together, raise the right leg straight upward to a forty-five degree angle with the floor, no more, no less. Do not turn or twist the raised leg and keep the right hip in contact with your right forearm. Toe of right leg should be pointed, knee locked, muscles tight.

Stay there ten seconds, 80–20 breathing.

2

The main pitfall here is the same one we encountered in the Standing Bow Pulling and Balancing Stick poses. Yoga Catch-23: the raising leg of a beginner is always followed by an attached hip. Why? Because raising the hip makes it easy to lift the leg. However, Yoga isn't interested in what is easy. So keep both hipbones touching your forearms.

As with the Standing Bow Pulling and

Balancing Stick, this is a straight up-and-down, forward-and-back pose, no ballet turnouts. The bottom of the raised foot and the back of the leg and knee move straight up toward the ceiling, while the leg remaining on the floor stays totally relaxed.

At the same time, feel as though someone has hooked your big toe to wild horses and they are pulling it through the back wall. In other words, **stretch,** not height, is the important thing.

You may well get cramps at first. Flex and wiggle the foot to relieve them.

3

Slowly lower your right leg to the floor. Still keeping your arms under your body, lift your left leg straight up without twisting or lifting left hip. Hold for ten seconds, 80–20 breathing, toe pointed, knee locked.

IDEAL
Salabhasana

3

If you aim for a slightly pigeon-toed feeling you will produce the perfect "straight up and down." A pigeon-toed feeling will also allow you to keep both hips on the arms more comfortably.

REALITY
Locust Pose

4

Slowly lower the left leg. Tilt your head down so that your lips are on the towel. With your arms in the same position, lock your knees, legs straight, toes pointed, every muscle of thighs and buttocks tight as a rock.

Take a big breath and raise both your legs and hips off the floor to your belly button. Breathe 80–20. Stay there ten honest counts.

4

The third part of the Locust is usually voted "Pet Hate Number One." To this I shall simply say that eventually it will be as easy as falling off a log. Until then, if you have been fussing about your aching elbows, lifting both legs and your hips off the floor will give you something new to complain about.

It is, of course, entirely possible that you won't be able to lift the legs off the floor at all. It is possible you won't even be able to figure out **how** to lift them. Your nerve-message system from brain to muscles may be in such a state of disuse that the brain has discarded its road maps and you can't find your way around your own body! (See also Step 2 of the Cobra.)

Don't give up hope. It's simply a matter of patience. Almost like a person recovering the use of limbs after paralysis or illness, you must keep at it until the brain-muscle linkage is reestablished and you can send messages to the correct muscles at will. The ideal way for those legs to

go up is for the muscles of the lower back and abdomen to **pick** them up. So talk to your belly as well as to your spine and lower back.

But don't be choosey the first few days, or even weeks, about how you get your legs up. Try pressing the floor hard with palms and arms, use the grimace on your face, mighty grunts—anything to lift your legs. Try lifting on exhalation instead of inhalation. Getting them totally off the towel by hook or by crook and holding them there for ten honest counts is the name of the game.

5

Lower both legs slowly with control, not dropping them. Pull your arms out from under the body and relax them at your sides, palms up. Turn your face to one side and relax on your stomach with eyes open for twenty seconds.

6

Repeat the three parts of the pose for ten seconds each. Then relax with arms at your sides and head turned to one side for twenty seconds. Keep your eyes open.

5

Remember, if you have cheated on the ten counts, you have cheated only yourself. And not collapsing out of the posture is an important part of developing spine strength. So, despite the temptation, don't drop your legs to the floor. Also, collapsing out of this one could put a dent in the floor.

6

I do have some good news for you. First off, your elbows won't hurt after about a week, and neither will your tennis elbow, if you happen to have had one. Second, your legs are always much higher in actuality than they feel to you. After a few weeks of practice, sneak a peek sideways into the mirror. You will probably be pleasantly surprised— and spurred on to even greater accomplishments!

Benefits

The Locust Pose has the same benefits as the Cobra, but it is even more potent in the cure of any back or spinal problem such as gout, slipped disc, and sciatica. It cures tennis elbow and is also excellent for firming buttocks and hips.

Classnotes from Hilda

One lazy Sunday I decided to just "flop through" my Yoga. It wasn't until after the first set of the Standing Head to Knee that it hit me. I realized that I had done the Standing Head to Knee **completely**—everything bowed, foot completely flexed, forehead easily touching the knee of the upraised leg . . . and I could have balanced there all day. I had almost forgotten I was balancing. In the Standing Head to Knee, of all things!

And the magic didn't fade on the second set. I went into the same kind of trance, with no pain, no sense of effort, no difficulties whatsoever. I felt as though I had left my body and was standing to one side and watching myself. I had, without knowing it, "emptied" my mind and body, just like Bikram said. And that emptiness was immediately filled with strength, with power. If I had ripped off my leotard, I'm sure there would have been a Wonder Woman suit underneath!

I know it sounds crazy, but I grew stronger and stronger with each pose. Then, immediately upon finishing, I walked to the phone, made a call, and calmly settled a matter that had been hanging fire for months. I just suddenly knew what I wanted and announced it with total confidence.

I haven't achieved exactly the same thing since that day, but shreds of it have clung. At least I know what that sense of control feels like, and I know it will happen again. I've got time to work on it. After all, compared to those ancient Yogis Bikram was telling us about, seventy-five is still wet behind the ears!

POORNA-SALABHASANA
Full Locust Pose

EIGHTEEN

"Archie, you are all the time coming down from the last part of the Locust before I say 'ten.'"

"I was up for two seconds before you even said 'one.'"

"Yes," says Sylvia, "and you don't know English very well. You started counting backward when you got to 'nine.'"

"My English one hundred percent perfect. And how many times I got to tell you, does not matter when you go up or if I recite the Address of Gettysboro while you are there. 'Ten' does not exist until I say it. If whole class would go up when I **say** to go up, I might be a nicer person. But I say '**Up**' and half the behinds still wiggling, and sissies crying about elbows, and grunting and groaning. Why I not have a class that looks like Rockettes? Instead I got Mexican jumping beans. The only time you allowed to cheat on counts is when you got a medical problem and I say you can take it easy. That is only time."

"Bikram," says Gail, "I was getting terrible cramps in my feet that time. What do I do about it?"

"There is nothing you can do about it, honey. Cramps comes and goes, even I sometimes get cramps. You just have to fight your way through them—shake out your foot, push hard with toes against the floor, do the best you can, and start again.

"Most important thing is, don't fright, don't scare when they come. It is a natural thing, and it is happy thing, for it means you are shaking up your body, getting it all disturbed and awake. It might be angry to be awakened, but pay no attention to it.

"Sad thing is when people let something like a cramp scare them. Some people always looking for an excuse to quit. Any little pain, any little twinge, they bring in the towel, run up the white flag, say to the body, 'Okay, you win.' Even sadder thing is, the body hasn't won when they do that, no one has won, except maybe the old-age doctors."

"Bikram," says Charlotte, "it's just amazing to me the resistance some people have to Yoga. I mean, I have friends who, when I tell them I am doing Yoga and benefiting from it, they get livid. They wonder if I'm turning hippy, tell me it's dangerous, that I'm not being fair to my family. There's just no talking to them. How can something so incredibly beneficial be so maligned?"

"And by people who don't know the first thing about it!" says Archie. "What about Yoga in India, Bikram, does everyone do it there?"

"What do they say—the shoemaker's child always has holes in the shoes? No, everyone in India doing Yoga no more than every American doing Jitterbug.

"But I am not unhappy about attitude toward Yoga in the West. You got to realize, world has changed a lot in last ten, twenty years. Things more and more understood and accepted now that people thought was pure crazy fifteen years ago, or ten. Can you imagine jogging being popular twenty-five years ago? Little by little, people finding out what you in this class already know. And pretty soon, people will be doing their Yoga every day like they already taking their vitamins.

"Okay, enough talking. Begin please. . . ."

1

Lying on your towel on your stomach, stretch the arms out to the sides, palms down. Put your chin on the towel and your knees, legs, and feet together. Point your toes. Tighten the muscles of the calves, thighs, and buttocks tight like rocks.

1

Some students are tempted to let their knees and their feet part company when straining during the second step of this pose. But if you keep all your muscles tight and toes sharply pointed from the start, this will be less of a possibility.

2

On a big inhalation of breath, all in one movement, look up at the ceiling and raise arms, head, chest, torso, and legs off the floor—arms tight and flaring backward like jet plane wings.

Palms should be facing toward floor, all fingers together, and arms and hands on level with tops of shoulders. Use back muscles and lower spine strength as much as possible.

All of your body lifts off the floor and arches more and more upward, like a beautiful flying bird—till you are balanced on the center of your abdomen only. Keep your feet and knees together and toes pointed, breathe 80–20, and stay like statue for ten honest seconds.

2

I want a 747, not a B-29. Get those arms up and back, hands always on a level with the shoulders. This means for each inch you lift your torso, the hands must lift the same distance.

The Full Locust is a subtle sneak. Remember I mentioned doing the Cobra by raising the torso without supporting yourself with hands or arms? Unmask a Full Locust and you find a snake in the grass.

And what's this business about raising your legs off the floor till you're balanced on your belly button? Didn't we just do that in the Locust, but with hands and arms to help keep us up and balance us? Now we do the same thing **without** benefit of arm strength, front or back.

There are no shortcuts or easier ways to develop this posture. Mastery takes sweat, strain, and determination. If sacrifices must be made (and as a beginner, the sacrifices are seemingly endless), sacrifice height in the legs in favor of getting the torso and arms arched up, and swept back just as high as possible. The ultimate will see legs and torso raised equally, but in the beginning stages accept the fact that the legs will lag behind. Just try honestly to get the legs off the floor a speck higher each day.

Other than that, have a good flight!

3

Come down slowly. Turn your head to one side, arms relaxed at sides, palms up, and rest quietly for twenty seconds. Then repeat the posture for ten seconds and again come slowly down and rest for twenty seconds.

3

To show you how subtly difficult the Full Locust is, it is the one position where I seldom play games with the ten seconds. If you are doing it honestly, giving it all you've got, ten seconds is all you can take. But oh! the wonderful things it does for your body.

Benefits

The Full Locust has the same therapeutic value as the Cobra Pose and the same upper-body benefits as the Standing Bow Pulling. It also firms abdominal muscles, upper arms, hips, and thighs.

Classnotes from Bertha

When I found out, ten years ago, that I had high blood pressure, my family settled upon my illness with a well-meaning frenzy. They were going to get me healthy. And their idea of how to get me healthy was to forbid me to do a thing. So for ten years I sat around like an invalid, waited on hand and foot. Nice, up to a point. But, naturally, I put on weight, got an even higher cholesterol, and my blood pressure climbed through the roof.

One day I decided that ten years of this pampered imprisonment had been enough. Now I was going to try to make me well in my own way. So, without telling anyone my plans, I bought a leotard and presented myself to Bikram.

After a week of Yoga I had my blood pressure checked. It was as high, if not higher, than ever. But when I told the doctor I'd started Yoga he said, "Wonderful, keep it up." I was mystified, but back I went for the second week—"under doctor's orders."

A week later, I again had my blood pressure checked. It was absolutely normal!

No one was more astonished than the doctor. "I'd **expected** results from the Yoga, but not this fast!" he said.

Now, after two months of normal blood pressure, with the weight falling off me and apples blooming in my cheeks, my family has finally stopped urging me to quit Yoga. They simply can't dispute the results!

The moral to this story is, it's **your** life, you live it.

DHANURASANA
Bow Pose

NINETEEN

"Bikram?" It's ten-year-old Barbie. "We went to Chicago to see my aunt last week. We went on a 747. And you know what? The wings go up and down. We flew through a storm, and we were bouncing all over, and I looked out and thought the wings were going to come off, because I could see them flapping."

"Sure, honey. If jet planes' wings didn't flap like birds', they would break right in two. The Empire State Building, you know, is built to sway many feet in the wind. If it did not have inner flexibility, it would come tumbling down. Same with big bridges. And a car with no springs or shock absorbers would rattle to pieces. So the health of even inanimal things depends upon balance, which is strength and flexibility in combination.

"Can you explain me why human beings recognize this for buildings and bridges and airplanes and cars but forget importance of keeping their own bodies flexible? Why they fail to see that they will break apart if they become rigid? And in human beings, this danger is even greater. Because man has a mind like no other thing has, and he allows the mind, as well as the body, to become stiff. And so he gets double stiff!

"All human beings should be like that jet plane, Barbie. Regularly it gets screws tightened, parts replaced, joints and hinges oiled, fuel checked to make sure is pure, all its electronic nerves tested. If plane was allowed to fly without this checking and oiling and replacing parts, it would get stiff and sick and pretty soon end up in junkyard."

"But Bikram," says Barbie, "you can't replace parts in a human being—at least not many, yet."

"That is not true. You are replacing your parts every day. There is no single part of you now—'cept brain—that was in you the day you were born. Every other cell you were born with, they died and were replaced by new ones. This will continue all the time you are 'young.' Old age is when the body has been so neglected that it can no longer replace its parts or when neglect has allowed your organs to become hopeless cases. But if you service yourself religiously like airplane company, I promise you are never going to go to junkyard.

"Practicing your Yoga every day is way to service yourself completely. If you freeze like rabbit when you have little cold or ache, if you scared to move or exercise because you have bad back or arthritis or high blood pressure, or if you fright to breathe if you got asthma or emphysema, if you sit with feet up when you get pregnant, or think way to fix bad complexion is to rub junk on your face, then you are terrorized by own body as though it is cursed tomb archaeologist is 'fraid to open, and you will suffer more."

"Bikram, I still don't really understand why something like jogging isn't as complete an exercise as Yoga. I always feel great after I jog."

"Then keep jogging. Jogging is nice thing. I have nothing against it, or other kinds of sports. Most are very good for Circulatory System, but health of human body depends on more than just Circulatory System. Depends upon systems like Abdominal, Spinal, Skeletal, Respiratory, and Nervous. These systems got to have more than just crumbs that Circulatory System throws them when it is being exercised, more than a little oxygen and a good shaking. They got to be specifically exercised.

"Jogging, for instance, is not exercising your all joints, where is hiding the calcium deposits causing arthritis, not squeezing and stretching your internal organs, glands, muscles, and tendons where are hiding hyperacidity, chronic digestion problems, hernia, kidney stones and other disease, gout, thyroid and tonsil trouble, appendicitis, problems of female organs, and ulcers. Jogging is also getting only a part of the lungs, very shallow. So is not really strengthening the lungs, making them elastic, and so is not really getting at the root of emphysema and asthma, bronchial and sinus problems, all the things to do with your breathing. It is doing nothing good for your spine, where is lurking lumbago, sciatica, slipped disc, rheumatism, and arthritis. It is doing not one thing for your Nervous System, where are hiding so many horrors, including loose screws, I could not name them all.

"All sports, all exercises, even something like ballet—it is same thing. Only Yoga maintains all the systems your health depends on. It is simple as that. You know who are healthiest people in world? My students who stick with Yoga.

"And so, begin please. . . ."

1

Lying on your stomach on the towel, bend your knees and bring your feet down toward your buttocks. Reach backward with your arms, take hold of each foot from the **outside,** grasping the insteps firmly about two inches below the pointed toes. Your hands and wrists should be on the outside of your feet, fingers and thumbs all together nicely. Keep feet and knees six inches apart throughout pose. Exhale breathing.

Note: Beginners with high blood pressure must never do this pose without a teacher's supervision. See medical appendix.

1

Once into this position you sometimes have the uncomfortable feeling that someone is going to come along, stuff an apple into your mouth, and start turning you on a spit. Indeed, the first few days some people can do nothing more for twenty seconds than lie there, clinging to their feet and looking helpless. (Which is at least better than the occasional person who can't even **reach** the feet.)

Be not discouraged. You have hordes of company. In my experience, this is the position that presents the most problems to the most people. It is where everyone's inflexibility really shows, and where we seem to have the most difficulty getting the messages from the brain to the muscles.

2

On a deep inhalation, look up at the ceiling and simultaneously lift thighs and upper body off the floor. Kick backward against your hands, lifting the legs ever higher off the floor.

Roll your body weight farther forward the higher you kick. The goal is to balance on the center of your abdomen. Once to your limit, breathe 80–20 and stay like a statue for twenty seconds.

2

It's fairly easy to lift the torso—indeed, in this trussed position the torso could hardly be anything but raised. It's those legs that sometimes refuse to budge off the floor, no matter how you kick backward or try to lift them.

To overcome that problem, first make sure that you have a good firm grip on your feet, then center your attention on the small of the back and your buttocks and forget that you are supposed to lift the torso and legs. Instead, imagine that you are going to push your abdomen, buttocks, and lower back right down through the floor. To do this you must make the muscles tight like rocks. Now push downward. More. Simultaneously press up and back against your hands with the tops of your feet just as hard as you can. **Up. Push. Hard!**

Your torso, by now, will be well lifted, and you should be feeling the first stirrings of lift in your thighs and will recognize the muscles you need to lift them—not in the legs but in the buttocks, lower back, and abdomen.

As an added tip, once you do get the legs up off the towel, become aware of how stiffly you are holding your shoulders. Let the arms pull the shoulder blades backward. You might even have to force the shoulders back the first few times to get the feel of it. But releasing the shoulders—and rolling your body weight forward on the abdomen—will allow everything to go higher, farther, and increase the stretch and benefit.

3

Lower the torso and legs slowly, turn head to the side and relax arms at sides, palms upward. Rest twenty seconds.

4

Repeat the pose for twenty seconds, then slowly lower torso and legs, and rest twenty seconds.

3

The Bow Pose combines the difficulties of the Cobra, Locust, Full Locust, Standing Bow Pulling, and Balancing Stick. But you are also combining all those wonderful benefits, which should make your rest sweeter.

4

If you can't go the full twenty counts the first few days, hold it for ten counts to begin and add two counts each day. A raw beginner's lack of endurance, bad cramping, high blood pressure, and **real** complaints as from a trick knee are, however, the only permissible excuses for quitting short of twenty counts. Besides, the Bow Pose is pretty once you get it. Doesn't that make you feel better?

Benefits

The Bow improves the functioning of the large and small intestines, the liver, kidneys, and spleen. It helps straighten rounded spines, relieves backaches, and improves pigeon chest by opening the rib cage, permitting maximum expansion of lungs and increased oxygen intake. The Bow also revitalizes all spinal nerves by increasing circulation to the spine. It improves digestion and strengthens abdominal muscles, upper arms, thighs and hips (it is especially good for increasing the flexibility of dancers' hip joints). It also improves the flexibility of the scapula, latissimus, deltoid, and trapezius muscles.

Classnotes from Francis

Hi. I'm the guy who had the operation on his knee, remember? A year ago I wouldn't have dared to open my mouth and give you classnotes. But now I have something to say. That's really what Yoga has done for me. It's given me a sort of quiet confidence. My knee has improved, sure, and it makes living a lot more pleasurable. But the **confidence** is really the thing.

I used to be sick a lot, and I lost a lot of time from work. Now I feel great and the tensions don't get to me. I look around me at work and see guys who are really great physical specimens (I'm built sort of like Woody Allen), and I see why they get sick. They're tight as bowstrings with strains and stresses, their personal lives suffer, and they try drinking to find the answers.

And here I am going in exactly the opposite direction—toward health, flexibility, and balance, toward relaxation and the ability to cope, toward real inner peace. I've gotten a raise and I'm being considered for a promotion. I even have a steady girlfriend.

The best thing of all is that I've actually gotten some of my co-workers to come here with me, and a couple of them have stayed. It gives me a good feeling that I've been able to share my discovery, to help them get all that I've gotten from Yoga.

Just think how Bikram must feel. He's helped tens of thousands.

SUPTA-VAJRASANA
Fixed Firm Pose

TWENTY

"That last time in the Bow was real good, Archie. You got your thighs completely off the floor."

"Naw . . . I'm never going to be able to do that one right."

"Yes you will, 'cause I am your teacher. I could teach the Statue of Liberty to do the Bow Pose. Why you always laugh? Soon you will be doing it seventy-five percent correct. And the day after you do, you will read in the newspapers that Manhattan collapsed into the harbor, 'cause all the people rushed to the tip to see the Statue of Liberty doing the Bow Pose out on her island."

"Holding her torch in her teeth?"

"I don't care where she holds her torch as long as she holds the pose for twenty honest counts.

"Barbie, you want to show Terry Two how to do full Bow Pose?"

Barbie reaches around and grasps her feet, lifts up front and back, and almost casually pulls her feet onto her shoulders.

"Can you believe that?" says Florette.

"You can believe it, 'cause pretty soon, if you continue to do your Yoga each day, in just the way I tell you to, you'll be doing it too.

"Why you always thinking negative thoughts, Florette? Archie says he will never do the Bow Pose right, and you get jealous of a little girl. Both of you already see miracles from your Yoga. Archie's slipped disc doesn't slip no more, and you got beautiful figure. How many miracles you have to see before you finally learn to have faith in what you can do?

"Again and again I tell you what you must gain to be successful, not only in the Yoga but in all of the life. To have worked the little miracles you have already worked in Yoga, you have already gained little bit self-discipline, determination, and concentration, or you would be in same condition you were when you first started class. But what happen to faith and patience? How you can stand there and tell me it is impossible for you to someday do Full Bow like Barbie? You cannot see what you have already achieved?

"Keys to success I keep telling you about, they are not some dusty old philosophy. They are twenty-four-karat gold, studded with diamonds and rubies—exciting and glamorous. And they all belong to you. All you need is believe in yourself.

"And so, like Phoenix rising from own ashes, come up from your face-down relaxation and begin please. . . ."

1

Kneel down Japanese-style on your towel, knees together and buttocks resting on your heels, soles of feet facing upward.

Keeping the knees together, move the feet apart just the width of your hips and sit down between your feet, your buttocks resting on the floor and the sides of your feet hugging the sides of the hips. Keep feet doubled back on floor throughout posture.

Put your hands behind you on your feet, palms on your toes, fingers facing outward and thumbs inward.

IDEAL
Supta-Vajrasana

1

Many people at first are unable even to sit on their heels. And then to get the buttocks to the floor between the heels can take weeks. But when finally they do, they feel as though they have conquered Everest. And in a way, they have. For once people prove to themselves they **can** do something, which only the day before seemed impossible—there is no stopping them.

To get yourself limbered up and hasten your progress (and ease the inevitable cramps in your knees and feet), you might practice this pose while watching television in the evening. Just get down on the floor Japanese-style, spread your feet apart, and bounce gently but persistently to stretch the muscles and tendons in your knees and feet and accustom them to their doubled-back position.

Some people—especially men—have great difficulty going anywhere in this posture with the knees together. If you find that this is the case with you, sit with the knees

a little apart at first, and then force them back together after you have become proficient.

As for putting your hands on your feet, if you cannot even get your buttocks down to your heels as yet, you won't be able to reach the toes. But try. It will increase the stretch.

REALITY
Fixed Firm Pose

2

Holding on to your feet, slowly bend your right elbow down to the floor behind you, then bend left elbow to the floor, so that your torso is leaning backward supported by the elbows.

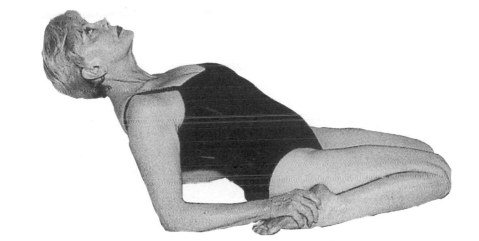

2

This portion of the pose usually brings out the first wail of protest from your feet. The farther back you go, the louder your little pigs will squeal, and the ankles, calves, knees, and thighs will take up the chorus. You have the whole-hearted sympathy of all who have gone before you, and profanities are allowed. Quitting, however, is illegal, immoral, and therefore out of the question.

But you can be confident of one important thing: The discomfort is occasioned by the newness of the position, not because you are injuring yourself. Only if you jerk into or out of the position too quickly is there the possibility of pulling something. Do everything slowly. That advice can't be repeated too often.

You may, as a beginner, put your hands flat on the floor behind you, rather than on the toes. This will allow you to get down onto the elbows with more security and to take some of the weight off your poor legs and feet and onto your arms and hands instead.

3

Keeping your knees together and touching the floor throughout, let your head drop back onto the floor and then relax your shoulders onto the floor allowing the elbows to slide out from under you after the shoulders and upper back are down.

Now raise your arms up over your head, opposite hands grasping opposite elbows, both arms flat on the floor. Tuck your chin down toward your chest. Completely relax, exhale breath, and stay there twenty seconds.

IDEAL
Supta-Vajrasana

3

There are two stages in this posture where almost everyone gets stuck. The first is at the point of getting both elbows lowered onto the floor. For a few days, you don't go any farther because you are legitimately stretching out all the muscles from chest through abdomen, hips, thighs, calves, and feet. But after that, freezing in this position is just plain **fear.**

Finally you summon your courage and actually drop your head back to the floor. Yet, there you freeze again, still bearing most of your weight on the elbows, not quite sure what to do next and afraid to try anyway.

The solution to both problems is to **relax.** Realize that by fighting the act of touching your shoulders to the floor, you are making it triply difficult for yourself! Let go completely. Let those elbows slide out from under you and relax the shoulders and upper back onto the towel. Your legs and feet will

complain, but bear it as long as possible and come up slowly.

Now that you are in the final position, begin trying to keep your knees absolutely together and flat on the floor. Then turn your attention to feeling as though your buttocks have turned to lead and are dropping right through the floor. Once you get the knack of that relaxation, you could fall asleep in the Fixed Firm, it's that comfortable.

But Yoga doesn't ask for heroes or fools. Do only as much as you can each day and then hold it there for the count.

REALITY
Fixed Firm Pose

4

Come up slowly, using elbows and hands for support. Turn around, lie down on your back, and do Dead Man Pose for twenty seconds.

5

Do a perfect sit-up, grasping toes and touching forehead to your knees. Turn around to face the mirror, sit down Japanese-style, and repeat Fixed Firm for twenty seconds.

6

Come up slowly using elbows and hands. Turn around, then lie down and do Dead Man for twenty seconds.

Benefits

The Fixed Firm Pose helps cure sciatica, gout, and rheumatism in the legs. It slims thighs, firms calf muscles, and strengthens the abdomen. It also strengthens and improves flexibility of lower spine, knees, and ankle joints.

More

More

4

No matter how desperately you may want out of the pose, you **must** come up slowly and in exact reversal of the way you went in.

5

Since the first set stretched your muscles nicely, you should be able to do it even better this time. Come on, don't be crybaby-chicken. Try.

When finally you reach the point of dropping off to sleep (this may take days, weeks, or months), then start moving the feet in closer to the hips and curling toes under the buttocks to increase the stretch on feet and legs.

When you master that, then sweet dreams. You deserve them.

6

Despite the discomfort of this pose, it's wonderful therapy to moan and groan and complain. The noises some people make should be recorded for posterity. And the axiom about misery loving company must first have been said about a class of Yoga students doing the Fixed Firm—though everyone always seems to end up laughing instead of crying.

ARDHA-KURMASANA
Half Tortoise Pose

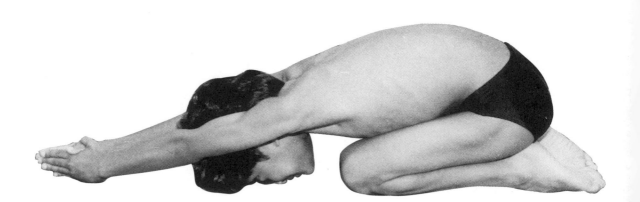

TWENTY-ONE

"Bikram, is the whole spine eventually supposed to lie flat on the floor in the Fixed Firm?"

"No, just the full shoulders and buttocks touching. It is almost physical impossibility for whole spine to touch floor in that one. Only person I ever saw could do it is teacher who works in one of my schools. Whole body is made of cooked spaghetti. She should leave it to medical science.

"Actually, it gives her big problem, though. First she must work very hard for strength to balance her flexibility. Second, she does not know what it feels like not to be flexible. So, she just says to students, 'Do this,' and she cannot really understand why they do not do it."

"It's like trying to understand what it is to be plain when you've always been beautiful," adds Florette.

"From which side of the fence do you assume yourself to be speaking?" asks Archie.

"It does not matter. She is now beautiful from doing the Yoga every day, and as long as she continue—or anyone continue—they going to stay that way."

"Bikram," says Hilda, "you started Yoga so young—four years old. Don't you find that you've forgotten what it was like at first? Or were you just naturally flexible?"

"No, I was not naturally flexible at all. I had to work very hard. But sometimes now I forget. And when I forget, then I make great sacrifice for my students. I pretend to be lazy for a month, I do no Yoga at all. Then, when I go back, I got to fight and hurt and be stiff and sore just like the rest of you."

"Gosh," says Florette. "How noble."

"Doesn't it just make you want to cry?" says Juliet. "Whenever it seems that Bikram is lazy and goofing off, he's really sacrificing himself for us."

"Okay, you guys, cut it up.

"Do a perfect sit-up, turn around on your towel to face mirror, and begin please. . . ."

1

Kneel down Japanese-style, sitting on your heels, knees and feet together, hands comfortably on your knees.

1

Those of you who had difficulty doing this in the Fixed Firm will have the same problem with the Half Tortoise—a gap between buttocks and heels. As suggested before, take every opportunity that presents itself to get down on your knees and bounce gently to stretch out your atrophied muscles, joints, and tendons. For the moment, however, sit down as far as you can and concentrate on keeping a downward pressure on your buttocks throughout the posture.

2

Raise your arms up over your head side-wise and make a nice steeple—palms together, thumbs crossed, arms touching ears, spine straight. Stretch your arms upward as much as possible and take a deep breath.

2

Half the benefit of this stretch comes from keeping your arms locked and touching ears as you bend forward in the next step. So, make sure you don't drop your steeple.

3

Keeping your buttocks touching your heels at all times, bend slowly forward from your lower spine in a nice straight line from tailbone to fingertips, keeping the steeple perfect, stretching the spine forward max-i-mum, and emphasizing big exhalations. Go down until you touch the sides of your hands on the floor, elbows locked, steeple still perfect, buttocks still resting on heels. Keep your eyes open.

IDEAL
Ardha-Kurmasana

3

The Half Tortoise feels like pure Heaven after the Purgatory of the Fixed Firm. Not that the Half Tortoise is simple, which fact you will grasp the moment you find that **no way** will your buttocks stay touching your heels as you bend forward. In all likelihood, the gap will be just as large once you get your forehead on the floor. Even Jeff still has an inch to go.

Next, you'll discover that you probably don't have enough spine strength to go all the way down with a perfectly straight spine. Resurrecting those crocodiles you met in the Balancing Stick in Chapter Seven will help; they're just where your face and hands will touch the towel. So it behooves you to go down slo-o-o-o-wly and to keep your face and hands off the towel till they absolutely must touch.

Not only will this slow stretch rapidly strengthen your back muscles and keep your spine as straight as possible at your particular point of development, but it will also keep your buttocks in contact with your heels longer. You'll learn the exact point at which they part company and be able to gauge your progress with each passing day.

REALITY
Half Tortoise Pose

4

Stretch your steeple forward more, as much as possible. Then touch your forehead to the towel, chin away from your chest. Totally relax your shoulders and back. Eyes still open, exhale breathing, stay for twenty counts.

Quincy Jones and Bikram

4

Quincy could be stretched even more. I sometimes help students stretch more by standing on their hips and bouncing gently while they are down there. That helps stretch spine and those muscles that need to stretch to get good buttock-to-heel contact. And it doesn't hurt. Matter of fact, it feels good.

And to get the best stretch with your steeple, as the sides of your hands (palm to palm, thumbs crossed) touch the towel, slide them forward. When they won't slide any more, then **walk** them forward with side-to-side, rocking-crawling movements, until your arms and shoulders are stretched to what seems the breaking point. Then lower your forehead to the towel—and relax.

Because relaxation is the aim of the posture, it works magic on tense neck and shoulders. See to their relaxation actively, chasing out any tenseness that persists.

(Once into your best stretch, and every day it will be farther, the stretch itself and your body weight do the exercise for you.)

5

Come up just exactly as you went down—slowly and in one solid piece. Keep hips touching with heels, steeple perfect, back straight. Then lower arms down sidewise.

Turn around and relax on your back in Dead Man Pose for twenty seconds.

6

Do a sit-up, grasping your toes and touching your forehead to the knees. Turn around to the front, kneel Japanese-style on your towel, and repeat the pose for twenty seconds.

Come back up just as you went down, arms down sidewise, turn around, and rest in the Dead Man Pose for twenty seconds.

IDEAL
Ardha-Kurmasana

5

It is easy to have buttock-to-heel contact by buckling in the middle as you lift up. It is not easy to keep that contact when you come up like one solid chunk of steel from fingertips to tailbone—but that is the correct method. And the only way you are going to do it is as slowly as a pregnant snail.

You must generate a tremendous amount of lift in your hands and arms and back muscles and keep that lift going all the way, or the arms will sag and you will buckle. When done properly, your whole spine, from top to bottom, will be working and strengthening. You'll feel it every step of the way, and it will be very satisfying.

6

Good thing you're all rested, 'cause the Camel is next.

REALITY
Half Tortoise Pose

Benefits

The Half Tortoise provides maximum relaxation. It cures indigestion and stretches the lower part of the lungs, increasing blood circulation to the brain. It firms the abdomen and thighs. And it increases the flexibility of hip joints, scapula, deltoids, triceps, and latissimus dorsi muscles.

Classnotes from Barbie

You know, I think I'm the most misunderstood person in this class. The grown-ups think just because I'm flexible, I don't work as hard as they do. They think kids don't really need Yoga, and they wonder what I'm doing here. They think I just like to show off, I guess.

Children are people, too, you know. A few of my friends are flexible like me, and can do a complete Locust, a Full Bow, and a Full Camel. But not many. The rest have just as much trouble as the grown-ups . . . and everyone likes to forget that I couldn't do it at first, either.

Charlie and Charlotte's son takes Yoga, too. He started out stiff as a board. Everyone said, "That's different. He's thirteen, he's almost an adult." That doesn't make any sense. People accept the fact that Tommy should be stiff because he's thirteen and almost grown up, and then they turn around and say that children don't need Yoga. If that's true, if I didn't do Yoga I would be as stiff as Tommy in three years. So why shouldn't I do Yoga and not get stiff in the first place?

I'm real proud of my Mom, too. She's pretty old—fifty-five. I came along late, I guess. I have a sister who is thirty. I can't remember having real fun with my mother before she started Yoga—she was always tired and cranky, because she had bursitis really badly. But she's so young and full of fun now that it's my **sister** who seems old. Though I think Marge will start Yoga soon 'cause yesterday she was walking down the street with my mother, and someone took them for sisters. Marge didn't laugh.

Anyway, **please** don't think Yoga's not for kids. **Encourage** them to do it now and they'll have it for life.

USTRASANA
Camel Pose

TWENTY-TWO

"Bikram, what does the Full Tortoise look like?" asks Archie.

"Looks like real Tortoise crawling through desert. Lavinia, I want to see you stretching your steeple much more forward from now on in Half Tortoise. Because besides maximum relaxation, that pose helps you stretch lower part of the lungs and increases the blood circulation to your brain so you don't get senile at sixty-five, okay?"

"Okay."

"You gonna come tomorrow and show me how good you can do it?"

"Gee, Bikram, I can't. I have to stay home and make cookies for the Girl Scouts."

"Cookies? Lavinia, you just said wrong word. Good thing you got twinkle in your eye. You make cookies tonight. So tomorrow you come to class and bring me some.

"Hey, Barbie, you want to show Terry Two what's a **Full** Camel?"

Barbie sighs. Then she stands up on her knees and arches backward all the way to the floor, doubling her head, shoulders, and arms under her torso so that they face forward through the knees. Her little body forms a perfect circle.

"Beautiful. Now stay there and I do Crow Pose on your ribcage. All set?

"One, two, three, four . . . how you feel, Barbie? . . . Seven, eight, nine, ten. Okay, that's enough! Look at her. Looks like puff

of wind could blow her away. But all she got now is pink face. Did you feel my weight, Barbie?"

"No."

"'Course not. None of rest of you has got to do that yet; that is **Full** Camel. But Barbie and I just showed you better than any of my words what strength there can be in flexibility, and that flexibility and strength combined make you balanced. When you are balanced, you can carry whole world on your shoulders and not even know it is there. That is what you are going to get from Hatha Yoga. There will be no weight, no pressure, no strain that life can put on you that you will not be able to support.

"Know what 'Hatha' means? It is Sanskrit. **Ha** means Sun and **Tha** means moon. Sun and Moon are the two sides, the right and left to the universe we know. So the ancient Yogis saw that there was also a right and a left in every part of human being—a strong side and a weaker side. And they named the right of the body Ha, meaning Sun, and the left of the body Tha, meaning Moon.

"Now we know how important is Sun. We get most of energy and nourishment from Sun. But Moon is not just decoration in sky over Lane for Lovers. Moon has very strong effect on all things of Earth—oceans, for instance. And on mind, too. Everyone know people crazy during full moon. So is very important for us to understand powers of

162

both Sun and Moon, and to balance them out.

"That is exactly what we do in Hatha Yoga. I told you Yoga means **union.** So, in Hatha Yoga physically we balance the right and left powers of body to make sure, for instance, nothing is crooked, not one hip higher than the other, one knee weaker, one side more flexible, or chronic pains in one side of neck or shoulder. When union or balance is achieved, all systems work in perfect synchronization and health—just as night and day, Sun and Moon, work together in Nature.

"And mentally you also balance, so that when Moon is full you do not grow hair and fangs and rush out like werewolf into night.

"Okay, let's begin please. . . ."

1

Stand up on your knees on your towel, knees and feet six inches apart. Put your hands on the backs of your hips, fingers pointing toward the floor. Inhale breathing.

IDEAL
Ustrasana

Freda Payne 1977

1

If it's more comfortable for your knees to be a little farther apart, that's okay. But keep your feet only six inches apart.

REALITY
Camel Pose

Freda Payne 2000

2

Keeping your hands on your hips, drop
your head back completely. Then bend the
torso backward slowly about six inches and
stop.

Freda Payne 1977

More

More

2

Even this much is sometimes tough for beginners. But those hands are on the backs
of the hips for good reason—for support. So
use them. And drop the head back **completely.**

Freda Payne 2000

3

Now, bring your right hand down and take a firm hold of your right heel, thumb on the outside, fingers pointing inward. Then bring your left hand down and take hold of left heel, thumb again outside and fingers pointing inward.

Take a deep breath and then exhale as you push your thighs, hips, and stomach forward as much as possible, using all the spine strength you have. Simultaneously, arch your torso backward max-i-mum. You should feel it in the small of the back. Emphasize your exhalations and hold pose for twenty seconds.

Freda Payne 1977

IDEAL
Ustrasana

3

The first day you will probably not be able to go any farther than gripping both heels. But it's a good beginning done correctly, so be patient with yourself.

One good thing about the triangle of pig iron you will resemble is that you can't get any worse.

And so . . . concentrate on the area from the top of your thighs to your waistline. Push it up and forward with everything you've got. Exhale and push harder. That mid-section will eventually begin to seem like an accordion, stretching with each exhalation.

When you have pushed forward as far as you can go that day, then change your focus of attention to the small of the back and try to release those tensed muscles. I say "try," because everyone is so convinced that he or she is going to break, that all the muscles fight the relaxation which **must** happen in that spot to enable the upper body to arch backward fully.

The day you finally gather your nerve and let the back release, the elation of feeling your body arch gracefully over will more than compensate for any discomfort.

Once you begin to get a good forward push, take care not to "cheat" by letting your hands creep upward off the heels. Keep your fingers well down into the insteps, grasping the heels fully and firmly.

REALITY
Camel Pose

Freda Payne 2000

4

Come up slowly the way you went down—right hand to right hip, left hand back to left hip to help you straighten. Turn around, lie down on your back, and relax in the Dead Man Pose for twenty seconds.

5

Sit up on a big inhalation, exhaling as you grasp your toes, laying your forehead on your knees and trying to touch your elbows to the floor as well.

Then turn around and do second set of the Camel for twenty seconds. Come up slowly, turn around, lie down, and relax in Dead Man for twenty seconds.

4

Some day I am going to snap a picture of the entire class emerging from the Camel Looks like a bunch of Zombies. For a split second I don't have any students in my class, they have all gone to a far land.

5

In my class, you know you are doing good Camel the day I stand on your hips while you are in the pose. When pushed fully forward, hips are like Rock of Gibraltar. You do not even know I am there. But, as you can see, there is always a "farther forward" and a "more back," until you are bent completely double.

Benefits

The Camel stretches the abdominal organs to the maximum and cures constipation. It also stretches the throat, thyroid gland, and parathyroids. Like the Bow Pose, it opens a narrow rib cage to give more space to the lungs. And because it produces maximum compression of the spine, it improves the flexibility of the neck and spine and relieves backache. It also firms and slims the abdomen and the waistline.

Classnotes from Leslie

If you think this is the pose that will break your camel's last straw, it's not. It only feels that way. Unfortunately it never stops feeling that way, no matter how advanced you get.

I must confess that in this one position, I have never conquered my fledgling fear that I would get stuck down there permanently. I think it's a holdover from a horror movie I saw at a very impressionable age. Jane Powell went onto a stage to dance with Ricardo Montalban; she was wearing a medical corset and it locked in a backbend. Shocking movie, called **Two Weeks with Love.** They shouldn't show such terrifying things to young people. Especially future Yoga students.

Of course there are some people (there's always a Barbie in every crowd) who can do a **Full** Camel. We normal folks ignore show-offs like that. (As a matter of fact, when Bikram isn't looking we pull the plugs on their heaters.)

Do I sound hard on the Camel? No better than it deserves. But I've been asked to say something nice about the Camel, so I'll say it. It's a wonderful thing to have just finished.

1977 **2000**

Emmy Cleaves and Bikram

SASANGASANA
Rabbit Pose

TWENTY-THREE

"Bikram, why is that called the Camel?"

"Why so interested suddenly in names of postures, Archie? Yoga is thousands of years old. Somewhere along the line a Yogi thought the position was like the hump of a Camel and named it the Camel."

"But it doesn't look anything like a Camel."

"So what? I am not responsible for a Yogi with bad eyesight. Come on now, you are supposed to be relaxing in Dead Man after your big effort, not doing research. Arms at sides, palms up, complete relaxation, let the floor support you. Keep your mind on your body, feel how the blood is going and where it is going. Feel each of your vertebrae as separate and apart from the others, and how very good they feel now to be freed from the Sheffield steel you had them cased in. Feel each separate muscle in the pelvis and abdomen and chest and neck, how alive each is after being stretched.

"You are **not** one great big mysterious chunk that will move in only narrow set directions and patterns and will rapidly get old and wear out. The body is capable of youth and vitality for your whole life, indefinitely, and of acting like a mix-and-match erector set. Parts can be separated, moved in different directions at same time, placed in most amazing oppositions and positions. Bones can change structure and direction at any age: move up, down, inward, out-ward. Ligaments, tendons, muscles can stretch like comic strip Rubber Man.

"Body is amazing apparatus. There is no limit to what it can do—and to what mind can do also. Constantly people ignore this, make the body a bad thing, a thing to be ignored, hidden, ashamed of, abused, **misused.** This is the greatest most single crime we can commit against the god, to treat stupidly the wonderful thing that has been given us.

"Many people think ignoring and hiding the body will help them to develop better their minds and spirits. Silly people. Brains cannot continue to work without good oxygen and nutrition. And you cannot concentrate on the mental or spiritual when your body is full of aches and pains and disease.

"My students have found a beautiful thing in Yoga. And will be beautiful because of it. Others will ask what you do to make you glow. And you can share what you have found so maybe they find new lives, too.

"This is what a Yogi does—shares with them his knowledge. Some of you already are Yogis. Others of you will become Yogis. But it makes me sad to know that one year from today, only twenty-five percent of you will still be applying what you have learned here. Five years later, only ten percent. But what good things that lasting ten percent will do for anyone who meets them, by being always an example with healthy glowing bodies and peaceful shining minds.

"It is not an easy path, to be a Yogi. Your Karma, your work in this life, is to help your fellows and to teach them, if only with example. And sometimes you will grow tired. A Yogi sometimes want to die—to spread the mat and go to sleep upon it. But it is not allowed. Yogi cannot die—does not **deserve** to die—until the Yogi has done all the good, helped all the people that the god intended to be helped. Only then can the grateful Yogi lie down upon the mat and close the eyes at last.

"Okay now, do better sit-up than ever, turn around, and begin please. . . ."

1

Kneel on your towel Japanese-style, knees and feet together, buttocks resting on your heels. Reach around and grip your heels so that your thumbs are on the outside and you are cradling your heels in your palms. Take a good firm hold with the fingers.

1

It is absolutely essential, for reasons that will be explained, that your grip is secure in this pose. To help you hold the heels securely (they can be slippery little devils), fold the edges of your towel over your heels and grasp the heels and the towel together.

2

Lower your chin to your chest and, with emphasis on exhalation, curl your torso slowly and tightly forward until your forehead touches your knees and the top of your head touches the floor. If there is a gap between forehead and knees, walk the knees forward until they touch the forehead, rather than reaching for the knees with the forehead.

2

This is a problem for most, but if you bend forward as you grip the heels, rather than sitting straight-spined, you'll find you are already halfway there. You have only to tuck that chin down into your chest and curl inward with a passion, as though reaching with the top of your head for the interior of a nautilus shell.

3

Simultaneously as you curl, lift your hips into the air, rolling your body forward like a wheel, and pull on your heels with all your strength, until your arms are completely straight and your thighs are perpendicular to the floor. Your feet stay on the floor. Your weight is supported by the tension between your arms and heels, not by your head. Put only twenty-five percent of your weight on your head.

Your neck may hurt a little, your throat should feel choked. There should be very little weight on your head. Breathe normally, eyes open, and stay like statue for twenty counts, exhale breathing.

3

Once you begin to lift the hips and pull hard with the arms, **don't let your hands slip off your heels.** If you feel your hands slipping, **immediately** lower your hips and decrease the pull.

The secret of the whole pose is in the arms. You must pull with all your strength on the heels of both to do the posture fully and to keep your weight where it belongs. But should you lose your grip, you could do an unexpected somersault and get yourself a stiff neck. So hang on.

If you find that you must walk the knees forward to meet your forehead, it shows that your back is not yet supple enough to stretch all the way. If you still cannot conquer the space between forehead and knees, time, patience, and persistence, my friend! If you feel dizzy the first day in this head-down position, you can hold it for ten seconds, then gradually work up to twenty.

But the choked feeling in your throat is no reason to stop. In fact, the more choked the better.

4

Still holding your heels, very slowly uncurl back up to your kneeling position, in exact reversal of the way you went into the posture. Turn around, lie down on your back, and relax in Dead Body Pose for twenty seconds.

5

Do a sit-up, turn around, and repeat pose for twenty seconds. Then turn around and relax on your back again for twenty seconds.

Benefits

The Rabbit produces the opposite effect of the Camel, giving maximum longitudinal extension of the spine. As a result, it stretches the spine to permit the nervous system to receive proper nutrition. It also maintains the mobility and elasticity of the spine and back muscles. The Rabbit improves digestion and helps cure colds, sinus problems, and chronic tonsilitis. And it has a wonderful effect on thyroid and parathyroid glands. The pose improves the flexibility of the scapula and the trapezius and helps children reach their full growth potential.

4

The Rabbit is neither exhausting nor painful and doesn't require great body strength or agility. You therefore don't have many excuses to avoid buckling right down and performing it properly. The object of the pose is to stretch the spine out slowly—as though the vertebrae were beads strung on elastic—nourishing everything on the band and aligning it, and then to release the tension slowly and let it all come together again. When stretched in a full Rabbit, it is not unusual for a spine to measure fourteen inches longer than usual.

Remember how the Camel compressed the spine? Well, this has the opposite effect. And the combination is pure magic for people with back problems. A Camel and a Rabbit a day generally keep the chiropractor away.

5

At this point you should feel like a million bucks, and for very good reason. It is not at all haphazard that the Rabbit is one of the last poses in my Beginner's Class. Up until now, you have been warming up so that you could give the Rabbit the fullest stretch you are capable of.

By the same token, it is important not to do the pose unless you **are** warm and **warmed up.** And if you have a fresh soreness or injury to your back, do not attempt to push anywhere close to your limit. Concentrate on slowly and gradually increasing the stretch, don't put too much weight on your head, don't lose that grip on your heels, and the Rabbit will soon become your favorite pet.

JANUSHIRASANA
Head to Knee Pose with
PASCHIMOTTHANASANA
Stretching Pose

TWENTY-FOUR

"Now is where all of you who have really been trying with your sit-ups will be rewarded. And I will give you another tip before we begin. You remember in the Pada-Hastasana, the part of the Half Moon where we bent forward? I told you to wiggle your hips to loosen the muscles and tendons so you could stretch down better. From now on, you do the hip-wiggling in your sit-ups as well.

"You do it in that split second between the big inhalation that gets you sitting upright and the big exhalation that helps you dive down and grab your toes and touch your forehead to the knees. Just in that second, with your torso bent slightly forward, wiggle your bottom like crazy and get all the flesh of your buttocks **behind** you, not in front of you or under you. Two things will happen. First, that wiggling will loosen and relax your muscles; second, you will gain precious inches for the stretch forward—and in Yoga, that can make the difference between doing the pose only eighty percent correct or one hundred percent correct."

"Kind of like a stage actress grandly sweeping her train out of the way," says Archie. "The ladies are to grandly wiggle their derrieres to the rear."

"Ladies? I don't remember saying **ladies.** What is that cushion **you** resting on, Archie?"

"Muscle. Solid muscle."

"Well, wiggle it to the rear, whatever it is. Understand, everybody? You will use that principle in all three parts of this pose. And when I say 'wiggle,' I want to feel this floor quiver like jello under your all bottoms.

"Okay . . . do beautiful sit-up from Dead Man, just like I told you, and turn around on your towel to face mirror. Ruth, you show class how perfect beginner can do first part of pose, and Dallas will do advanced big stretch, head to toes, in Paschimotthanasana. Begin please. . . ."

1

Sit on the floor and stretch your right leg to the right at a forty-five degree angle. Bend your left knee, bringing the foot up so that the heel is snugly against the crotch and the ball of your foot is fixed against the inside of the right thigh. Right leg stays absolutely straight with the knee locked, foot flexed back toward you. Now wiggle your hips and "put it all behind you."

Raise your arms up over your head sidewise.

IDEAL
Janushirasana with Paschimotthanasana

1

Some people's legs don't seem to bend this way at first. Just do your best. Soon you'll discover what "inflexibility" **really** means.

REALITY
Head to Knee Pose with Stretching Pose

2

Arms and head together, stretch down over the straight right leg, wiggling your hips more as you do. Take hold of your right foot with both hands—fingers tightly interlocked under the toes and the thumbs crossed on top of the toes.

2

The first fact that will strike most of you is that either your arms are too short or your legs are too long. Whichever it is, wiggling or no wiggling, you cannot reach your toes. You probably think I am insane to imagine you **could** reach your toes without bending your legs. The fact that Ruth seems to be able to do it makes no difference whatsoever to you.

So, as any level-headed person would do (indeed, you have my full permission), you bend the knee of your extended right leg upward until you **can** grasp the toes.

3

Now pull the toes back toward you as much as possible—this is a pulling exercise—and slowly bend your elbows straight down toward the floor. Tuck your chin to your chest and, without bending the right knee, slowly go down and touch your forehead to the knee. If your knee is bent, try to push it down with your forehead.

Drop or roll your left elbow down closer to the floor by rolling your body slightly to the left of your extended leg. Pull back **harder** on the toes—your goal is to lift the heel right off the floor. Eyes open, exhale breathing, stay ten seconds.

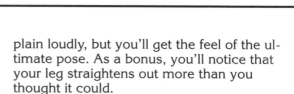

3

Remember my discussion of the Standing Head to Knee Pose? Well, as with that pose, kicking forward through the heel while pulling the toes back toward you, causing the leg to bow downward, is exactly what you are going to do here. If the leg were on the floor in the Standing Head to Knee (instead of extended in the air), pulling your toes back toward you and causing the leg to bow downward would lift the heel off the floor.

The nice thing about this sitting version of the Standing Head to Knee Pose is that you don't have to balance on one leg! (There. I knew I'd find some way to make you smile.)

Anyway, to prove to yourself that the heel can and eventually will lift off the floor (even if your leg is bent right up to your nose in order to allow you to grasp the toes), take hold of those toes and pull back on them with all your strength until the heel lifts. Your sciatic nerve will no doubt com-

plain loudly, but you'll get the feel of the ultimate pose. As a bonus, you'll notice that your leg straightens out more than you thought it could.

Furthermore, don't forget to use your forehead to push the knee down little by little, make sure you're bending your elbows toward the floor, and have faith. Pretty soon you'll be doing the pose with your leg straight.

4

Come slowly up and reverse the pose on the left side. Extend the left leg, bend the right knee, heel to crotch and foot fixed against inner thigh. Raise arms sidewise over head and bend torso forward from the hip, grasping the toes of the left foot with both hands.

Touch forehead to the knee and bend your elbows toward the floor for ten counts, eyes open and exhale breathing.

4

The inequality of your sides will be very apparent here. The left will either be much more difficult or much easier.

5

Slowly come up. Extend both legs straight out in front of you. You are now going to do Paschimotthanasana, the Stretching Pose.

In one fluid movement, using the momentum to help you stretch, lie back, raising both arms over your head, and do immediate sit-up on a big inhalation. Wiggle your hips vigorously as you begin to exhale, and stretch down over your straight legs. Take hold of your big toes with the first two fingers of each hand.

5

Now do you see how important the sit-ups are? A sit-up done correctly is half the battle won. Which ought to take the boogeyman out of the pose.

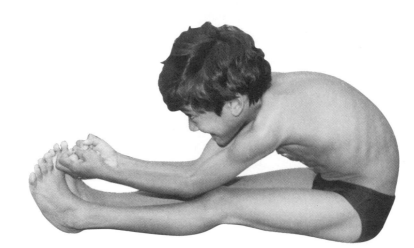

6

Pull the toes toward you as much as possible. Wiggle your hips right and left a few times. Now, with emphasis on exhalation, without bending your knees, touch your elbows to the floor, and touch stomach, chest, and face to your legs. Wiggle again. Your goal is to pull your heels off the floor and touch your forehead to your toes.

Stay twenty seconds, exhale breathing.

6

You'll notice, of course, that this is really a wee bit harder than a normal sit-up. In the sit-up you dart down for the touch and hold it for a few seconds, but in the Stretching Pose you must stay there for twenty seconds.

As a beginner, it will probably be impossible for you to touch your elbows to the floor and lay your body out along your straight legs—much less touch your forehead to your toes! Let's be realistic: Bending the legs will help some. Then pull hard on the toes while releasing them back toward you and pushing toward the mirror through the heels. Put your forehead to your bent knees and try to straighten them downward. Make it pull behind the knees. You **must** stretch those sciatic nerves.

In addition, everything I have said about mastering the Head to Knee Pose also applies to this posture. Your progress in one should parallel your progress in the other.

7

Come up slowly, turn around, lie down on your back, and rest in Dead Man Pose for twenty seconds.

8

Do a perfect sit-up, turn around, and do a second set of the Head to Knee to each side for ten seconds each, then the Stretching Pose for twenty seconds. Straighten slowly, turn around, lie down on your back, and rest again in Dead Man for twenty seconds.

7

This gives your screaming sciatic nerves time to quiet down before you once again put them on the rack.

8

There are many ways to attack both parts of this pose, many tricks to use to limber yourself and hasten flexibility. Play around with it. Don't be afraid, you won't hurt yourself. At any spare moment, get yourself warmed up, then start making your rocking dives toe-ward. Try lifting your heels, try reaching with your forehead. And remember to wiggle your hips and get your weight forward. **Go** for it, and you will make astounding strides.

Benefits

The Head to Knee Pose helps to balance the blood sugar level. It improves the flexibility of the sciatic nerves, ankles, knees, and hip joints; improves digestion; enhances the proper functioning of the kidneys; and expands the solar plexus.

The Stretching Pose relieves chronic diarrhea by improving the circulation of the bowels. It also increases circulation to the liver and spleen and improves digestion. It increases the flexibility of the trapezius, deltoid, erectus femoris, and biceps muscles, sciatic nerves, tendons, hip joints, and the last five vertebrae of the spine.

Classnotes from Charlie

The first part of this posture, the Head to Knee Pose, is another "TV Special." Sit on the floor and just practice rock, rock, rocking toward your toes, keeping the extended leg straight. Suddenly, in the middle of a **Casablanca** rerun you'll have hold of your toes. And you'll know what the ultimate posture will feel like.

That is the single most important thing that can happen to you, Terry Two—getting the feeling of the ultimate posture and the internal knowledge that **you can do it.** Once you have that, your awe of the posture disappears and you can then proceed systematically and logically.

As to the Stretching Pose, again and again you've been told that real breakthroughs come from relaxation. That's the way it was for me the first time I really did the Stretching Pose. I had "walked through" the class, feeling lazy and not at all like trying hard. When we arrived at the last part of this pose, I just flopped forward, my mind dwelling on the happy fact that the class was almost over. Then I realized that my elbows were resting with no strain on the floor, and my body was laid out perfectly flat on perfectly straight legs.

Unfortunately, Bikram realized it at the same moment. He raced over and jumped on my back, stretching the muscles and tendons even more with his weight.

"How's that feel? Good?" he asked delightedly.

"Yaaargghhwonderfullloooowwwhheee!"

"Pull more. Touch forehead to toes!"

I'm sad to say that despite his tender ministrations, I couldn't touch my head to my toes that day, nor have I yet. But I will.

"Nowhere in world," says Bikram, "will you see anyone who can do this impossible thing. Only **my students.** Because my students are the **best.**"

None of us will argue him on that point.

ARDHA-MATSYENDRASANA
Spine Twisting Pose

TWENTY-FIVE

"Hey, Archie, how come you didn't ask me why the Rabbit is called the Rabbit?"

"Oh," says Archie, "that one I can figure out. The feet are like the ears, and the body is . . ."

"Wrong. It is called the Rabbit in honor of cowards like you who won't pull hard on those heels and get the arms completely straight."

"I'm so glad I asked," says Archie.

"By same token, last combination—Head to Knee and Stretching Pose—should be called the Chicken and the Fraidy Cat. My goodness, I never heard so much clucking and yowling in all my life. There is not one of you stretching the sciatic nerve half as much as you could. But you are all putting on big act, ooooing and owwwing and making awful faces, thinking you fool me. You don't fool **me**, you only fool **yourself** when you are not one hundred percent honest with a pose.

"Only one really trying was Terry Two. And just for that, Terry Two, I have good news for you. This is our last pose. Just little breathing exercise after that. Oh, look that happy smile. I hope you are smiling just as happy when you try to get out of bed in the morning. But however you feel, you **must** do class tomorrow, just like Lavinia is going to come to class and bring me some Girl Scout cookies. Right, Lavinia?"

"Right."

"What? My ears must be playing tricks. You serious, Lavinia? You mean I did not talk myself hoarse for nothing?"

"I can't let you starve," says Lavinia.

"Oh my gosh. I got to give you something just for that. What I got for present? Pillows? Some candy? Candy not good for you. I think I give you box candy come in."

"Terrific," says Florette. "She can have it bronzed and put it on her mantle.""

"I just might do that," says Lavinia.

"Okay. Sit up on big inhalation, wiggle hips, exhale, and dive for your feet. Turn around and begin please. . . ."

1

Sit on the floor with both legs in front of you. Without allowing either buttock to lift off the floor, bend your left leg so that your knee is on the floor and your left heel is touching the side of your right buttock.

Now bend your right knee and bring your right leg over the left leg, putting your right foot down just to the left of your left knee. Your right heel should touch your left knee. Exhale breathing.

IDEAL
Ardha-Matsyendrasana

1

Remember those air-filled toy clowns with the weights in the bottom? You knocked them down and they popped right back up. Well, in this pose most people feel like a toy clown with the weight in the top. Set them up, and they fall right over backward.

First, you may have difficulty wrestling the curious appendages you think of as legs into the position described in the Ideal. Once into position, the knee on the floor lifts slowly and steadily, the foot thrown over that knee is carried along, and there you are—the Amazing Tumbling-Over-Backward Toy.

The key is: The hips **must** stay flat on the floor to do this position properly. It's the only way to keep your weight forward. But you have, in all probability, already allowed the foot curled around by the side of the right buttock to creep **under** that buttock, thus tilting you slightly backward. So from the beginning, put that foot farther out to the side than you think you should, leaving

a gap between the heel and the side of the right buttock of about three inches. Then press the buttock to the floor till it touches the heel, as opposed to bringing the heel up to meet the hip.

If you still have trouble, then simply rock forward and to the side, repeatedly forcing the right buttock closer to the left heel on each rock, until you stretch the necessary muscles and can touch comfortably.

The second essential part of maintaining the proper weight distribution is to keep the knee on the floor. It must not raise. If you have both buttocks flat, it will be much easier to keep the knee down.

REALITY
Spine Twisting Pose

2

Bring your left arm to the right and over your right knee. Press the elbow of that arm back against the right knee, and slip your hand in between the left knee and the right ankle, grasping the kneecap firmly with your palm.

Now put your right arm behind your back and reach all the way around your body until you can touch or grasp the left thigh.

Turn your head to the right and twist your face, shoulders, and whole torso to the right as much as possible. Both buttocks and left knee are still flat on the floor, and your spine is absolutely straight.

Exhale breath and hold for twenty seconds.

2

If you can't get your left hand in between the left knee and right ankle and still stay upright, grasp the left leg at any point where you are able—lower knee, shin— even the towel you are sitting on if necessary.

You may also use the right hand to prop yourself up, lifting it to reach around toward the left thigh only when you have sufficient balance.

You'll feel even more like a backward-tumbling toy at this point, but if you've done your groundwork properly you can fight the tendency. And in a surprisingly short time your muscles will strengthen and balance won't be a problem. You can then devote yourself to twisting with full gusto.

Normal breathing helps the twisting to no end. On each exhalation, twist around just a fraction more. Exhale loudly, that helps, too. For opposition in the twists, use that straight left arm to press your knee farther and farther backward as you twist more and more to the right. And if you can reach all the way around to grasp the opposite thigh, really take hold of it and use it to pull yourself around.

3

Unwind, reverse legs and arms, and do the pose twenty seconds to the left side. Twist your face, shoulders, and torso to the left as much as possible.

4

Unwind, turn around, lie down, and do a nice Dead Body Pose for twenty seconds.

IDEAL
Ardha-Matsyendrasana

3

For a while, you will be twisting the torso from the waist up. But once your spine is flexible and straight enough to twist the head and torso all the way to the back, then be sure to twist the abdomen as well. To do this, stretch your upper body toward the ceiling and lift the abdomen up and out of the pelvis.

Remember, every fraction of an inch counts in Yoga and sometimes makes the difference between perfection and just eighty percent correct.

4

We do not do this pose a second time. You'll just have to wait till tomorrow.

REALITY
Spine Twisting Pose

Benefits

The Spine Twisting Pose is the only exercise that twists the spine from top to bottom at the same time. As a result, it increases circulation and nutrition to spinal nerves, veins, and tissues, and improves spinal elasticity and flexibility and the flexibility of the hip joints. It helps cure lumbago and rheumatism of the spine, improves digestion, removes flatulence from the intestines, and firms the abdomen, thighs, and buttocks.

Classnotes from Reggie

Does this posture adjust the lower back! As a former sufferer from chronic back pain, I am here to assure you that the pose is all it's cracked up to be. But you'll eventually come to realize that no single pose or even group of poses does the job thoroughly. It is the **complete set** of exercises we have just gone through that puts everything into proper alignment. And even a half set—each pose done once—performed religiously each day will keep the spine, neck, and shoulders forever pain and problem free. As Bikram says, it's the scientific building of each posture on the last that results in perfect body balance and well-being.

I have yet to talk with anyone who has applied him- or herself honestly to this regimen who does not say exactly the same thing. The comments range from "It's a miracle!" to "I've just never felt or looked better in my life."

But if you've come this far and are still with us, you undoubtedly feel the same way and have your own success story to tell. You are one of those fortunates on whom the shot "took," Terry Two. Congratulations.

KAPALBHATI in VAJRASANA
Blowing in Firm Pose

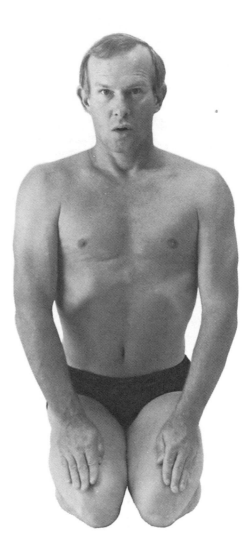

Tom Smothers

TWENTY-SIX

"That Spinal Twist is my favorite exercise to teach. Class is like big bowl of Rice Krispies, all going snap, crackle, pop."

"If he starts pouring milk and sugar on us," says Florette, "watch out."

"I already put the milk and sugar on you. You know how Rice Krispies is without milk and sugar?"

"Pretty awful."

"Just like you, when you first come to class—stiff, starchy, would crumble to pieces at the slightest pressure. Now is metamorphosis, from dry grain of rice to sweet, succulent morsel.

"Doing this Yoga you are changing your entire body, from internal organs to bone to skin, head to toe, internally and externally. This means whole body is functioning properly—which means you have strength, flexibility, and balance—which means you are synchronizing mental and physical powers to perfection. When all of this has happened, then, very easily and comfortably you will be a success in anything you set your mind to. And for the first time in this life you will know the happiness.

"So, my butterflies, sit up on inhalation, touch toes, and turn around. Don't look so worried, Terry Two. We started with a breathing exercise, so we end with a breathing exercise. Begin please. . . ."

1

Kneel down Japanese-style, spine very straight, hands resting on your knees.

Begin to blow your breath vigorously through your lips, as though blowing out a candle. Concentrate on the exhalation. Don't worry about the inhalation; it will happen automatically.

Simultaneous with each exhalation, pull your stomach in firmly. Immediately relax the stomach and contract it firmly with the next vigorous exhalation.

Repeat this sixty times slowly and rhythmically like a metronome, giving each exhalation one count. Don't speed up. Rest for several seconds in your sitting position and repeat another sixty times.

Tom Smothers

IDEAL
Kapalbhati in Vajrasana

1

The usual problem in this exercise is coordination. Again, it's like patting your head while rubbing your stomach. Pulling the stomach in while blowing out will at first seem contrary to what is natural.

Pretend there is a candle about a foot in front of you. Put one hand on your abdomen just above your waist, then vigorously blow that candle out. Did you feel the midsection move inward? Contracting the stomach muscles is the **only** way to force a strong enough gust of air out of your lungs to extinguish the candle. That exhalation, with the accompanying inward movement of the abdomen, is all you are being asked to do.

But oh, the comical gyrations your stomach will go through as it tries to move in the right direction at the right time. So to help things along, start with the stomach completely relaxed and hanging out. Come on, this is no beauty contest. Let every muscle go. You may have to **push** it out to get it to

sag if you've dedicated your life to holding it in.

Now look down at your midsection so you can see when you're getting the proper coordination. Blow out the candle and watch it contract. Don't blow out another candle till it has returned to its extended state. Adjust your rhythm to about one exhalation per second—or the time it takes to move your "bellows" in and then to let it relax.

The only other catch to the Blowing Exercise is that those stomach muscles are the **only** part of you that is supposed to move— no shoulders, arms, or lower back jumping in sympathy with the abdomen.

You'll get the hang of it quickly enough. And you're forcing every last drop of carbon dioxide out of your lungs, making room for fresh oxygen and improving elasticity. Your body isn't used to such good treatment!

REALITY
Blowing in Firm Pose

2

Turn around, lie down, and rest in the Dead Man as long as you like.

Benefits

This last breathing exercise strengthens all the abdominal organs and increases the circulation. It also makes the abdominal wall strong and trims the waistline.

2

You may even close your eyes now. You have earned sweet dreams. But don't miss class tomorrow.

Classnotes from All of Us

Abandon all preconceptions, all ye who enter Yoga. And go forth and conquer. Hang onto stair railings just in case your ascent or descent gets a little too bouncy, and grab the nearest lamppost if your feet refuse to stay on the ground.

And if the things happening inside of you communicate themselves to others, don't be afraid to share them. Not only might you have the joy of giving new life to others— you may even recruit some dependable classmates with whom to do your Yoga. And dependable is the key word.

In truth, though, you must depend only upon yourself. Impose your own self-discipline and bask in the glow Bikram's Yoga has given you.

Welcome to your new life, Terry Two.

Maintaining Proper Yoga Practice

After reading my book, I'm sure you understand that you will get from me no quick-frozen, powdered, instant, or irresponsible recipes for "total physical fitness" in fifteen minutes a week or even in fifteen minutes a day. Because there is no such thing. Instead, you will get from me a one hundred percent program for one hundred percent physical and mental well-being. My **Beginning Yoga Class** treats the whole machine for whole health, whole cure of chronic illness, realizing that the chronic weakness or malfunctioning of one little organ, joint, or vertebra can put the sleekest of human vehicles into the junk yard.

I cannot state strongly enough that no special set or specific recommendation for back problems or any other complaint changes the fact that the only safe and lasting way to cure, or relieve the symptoms of, chronic ailments and achieve total health is to perform the twenty-six poses I have given you exactly in the order and in the manner described, and on a regular basis.

With that caution in mind, below you will find guidelines about a special set, which you may try as you progress in your Yoga, and some words of advice about Yoga for children, bed patients, and the elderly.

FULL CLASS

Stage One. As a beginning, do the Full Class religiously each day for two months until you are performing all but the very difficult poses such as the Standing Bow Pulling and Stretching Pose eighty percent correctly. If you are restricted by any medical condition or have any chronic disease, you should continue daily until your condition is resolved.

Stage Two. Only then—when your progress is measured in fractions of inches that add excellence to what is already good, rather than in bold strokes, like being able to balance on one leg for ten seconds. Even then, you should continue your Yoga practice daily.

Stage Three. You are now performing ninety percent of the poses ninety percent correctly. Still you should not feel complacent about doing your Yoga. Doing this Hatha Yoga class every day is like everything that you do in life that is good to maintain your life. You brush your teeth every day until the day you die because you know that it is good for your health and it makes your mouth feel better. You do certain things every day of your life because they are good for you. And you don't plan to give up doing those things because of the daily benefit you receive.

Your body is telling you better than your mind that this Yoga practice is good for you. So listen to your body and give it the Yoga exercise that it craves each day. One you are hooked, you are hooked. And Yoga is a **good** habit.

How good is this Yoga habit? The pictures of Freda Payne and Emmy Cleaves in Camel Pose and Irene Tsu in Standing Head to Knee Pose show the results of daily Yoga practice of my Beginning Yoga Class. Twenty-two years between the photos. Don't ask their ages. As you can see, they have become ageless.

HALF CLASS

Stage Three Students Only. You may have to travel or be on an exceptionally busy schedule where you cannot do your regular Yoga practice. The Half Class guarantees that you will not lose the cumulative benefits that you have earned with your daily Yoga practice.

The Half Class is simply each of the poses done once in the prescribed order. Use it only when you are truly pressed for time. Since you're depriving your body of the second set, in which real progress occurs, a Half Class, when done *occasionally,* is really only "maintenance."

THE PICK-ME-UP SET

Stage Three Students Only. Muscles and ligaments of Stage One and Stage Two students are not yet strong or pliable enough to skip around, to mix and match exercises. Even for Stage Three students, the Pick-Me-Up is to be done only **in addition to** your regular Yoga.

This set works on problems such as these: low energy and midday fatigue; tenseness and irritability; backache; stiff neck; and headache. Or else just the need for a micro-mini vacation before facing it all again. Make sure you follow the suggestions explicitly, since the alternation of forward-bending and backward-bending poses is essential. Perform the following:

1. Standing Deep Breathing
2. Half Moon and Hands to Feet poses
3. A forward-bending position such as the Standing Head to Knee or Standing Separate Leg Stretching Pose
4. A backward-bending pose such as the Cobra Pose, Locust, or Full Locust
5. Triangle Pose
6. A second forward-bending pose such as the Half Tortoise or Rabbit Pose
7. A second backward-bending pose such as the Bow Pose or Camel
8. Spine Twisting Pose

Caution: Never do the Pick-Me-Up Set in a cold room, and do not force the muscles as hard as you would in a full series.

CHILDREN'S SET

Generally, a child's attention won't last through a full set of the Beginners' Series. But a Half Class, done as a game as often as possible, will form a good habit that the child will maintain for life.

Children with glandular upsets should be especially encouraged to do Yoga. As a tip to young boys, the Rabbit Pose is amazing in its ability to promote growth.

BED PATIENTS

Mobile bed patients can hasten their recovery by performing many of the poses in this book. For example, since digestion is often a big problem with bed patients, doing the Wind Removing Pose at least twice a day is ideal. The Dead Body Pose is a natural, and the Tree Pose and the Eagle can both be practiced lying prone.

Use your imagination. You could even do a version of the Standing Separate Leg Stretching sitting up in bed. And if you are able to lie on your stomach, try the Cobra, Locust, Full Locust, and Bow poses.

Bed patients should, though, use caution with the Head to Knee and Stretching poses (see "Constipation, Diarrhea" entries in the Medical Advisory section).

THE OLD AND INFIRM

There is no such thing as old and infirm. You aren't old, you've just been lazy for the last 200 years. I will tell you the same thing even if you tell me you are 101. Put on your leotard and get to work. Full class every day for two months minimum. Then you will see how silly you were to think you were old.

Medical Advisory

Yoga is not only good medicine, it's also excellent preventive medicine. By looking through this Medical Advisory section, you'll not only learn how a particular condition can be prevented, but—more importantly—you'll begin to understand just how much in command of your own health you are. From the experience of teaching over a million and a half students, I can confidently say that my system of Hatha Yoga is capable of helping you avoid, correct, cure, heal, or at least alleviate the symptoms of almost any illness or injury. That knowledge alone should make you feel better.

I've also included this medical advisory section, of course, to suggest how you can help yourself if you have any of the health problems listed here and to keep you from doing yourself an injury.

ANEMIA

Anemic persons should work especially hard on all lung-expanding and lung-stretching exercises. In addition to the full series, concentrate on the Standing Deep Breathing, the Blowing in Firm Pose, all back-bending poses, and the Half Tortoise. (**See also** Breathing Problems.)

ARTHRITIS, RHEUMATISM, GOUT

"Hopeless"-type maladies? Diseases of laziness are all they are. Yoga can "cure" arthritis. That is, it can relieve symptoms. This is not a miracle; it is common sense.

Many people think arthritis occurs because of an overabundance of calcium in the body. But there is really no overabundance. The problem is that the calcium is deposited as a form of calcium phosphate in the joint-tissue, including the spine. At that point, the calcium phosphate deposit begins to build layers in the joint—spiky crystal formations like a cactus—until no room is left for the joint to pivot smoothly in its socket. And these spiny needles irritate the surrounding muscles and nerves, and the agony of arthritis begins.

Rheumatism? It is closely related to arthritis, but is even more exclusively a lazy-person's disease. You have only to do your Yoga and you will be free of rheumatism.

Gout is also a problem that attacks the joints. And again and again in my series of exercises, you find me addressing myself specifically to exercising the joints. If I seem to be reducing some of the oldest, most painful, and perplexing diseases to lack of exercise, you're right. But that is what they often seem to boil down to.

Sadly, the theory seems to be that with advancing "age" one should "slow down," "take it easy," don't exert oneself or do too much exercise. And if you get something like arthritis, take it even easier, don't **move,** except to open your mouth to swallow the latest pill being offered as a cure. This advice is simply more nails for an earlier coffin. Exercise—meaning daily Yoga—is the cure.

BACK PROBLEMS: Including Stiff Neck, Whiplash, Frozen Shoulder, and Headaches

Picture your spine as a series of ball bearings (vertebrae) one on top of the other, each separated from the next by a cushion (a cartilage disc). When the spine is shiny and new, all the ball bearings are smooth and round, moving freely in all directions, and the cushions are strong and thick. Now picture your daily activities. In one position after another, probably ninety-five percent of the time, that spine is leaning forward.

What is happening, then, is that each vertebra of your back is compressing its cushion in a frontwise direction. This goes on year after year until there is no resiliency left in the front of the cushions, while the two sides and back have grown weak and slack from disuse. In addition, lack of movement has made the bearings rusty and barnacles have developed. The result:

backache, stiff neck, headache, and countless other complaints.

The cure: exercise! Make the spine work so that resiliency and strength are restored to each cushion, so that the rust and barnacles are worn off the ball bearings, so that an X ray would show them smooth and round, sitting snugly on their fat, renewed cushions.

Beginning with the Half Moon, my series of exercises is designed to make your shocked and shriveled spine work to both sides, to the back, and then to the front. Only by exercising in all directions can your spine be healthy; and only with a healthy spine can you have a healthy nervous system.

If your chronic problem is something such as sciatic pain, lumbago, sore back muscles, whiplash, vertebrae out of line, shoulder trouble, radiating pains down the arms, tension headaches, arthritis, or rheumatism, hunchback, swayback, spinal curvature, pinched nerves, or "something not quite right that the doctor said I ought to watch," stop watching. Act! Get to work on these exercises. Even those who have had spinal surgery should get to work—with their doctor's okay and a qualified teacher who can lead them in my particular series of exercises.

People with slipped disc are often in such pain that Yoga seems further torture. However, in numerous slipped disc cases, determined Yoga can save the day. So endure the pain. But please note that those with slipped disc should also work under the supervision of a qualified instructor using my exact series of exercises and the safety rules laid down in the body of this book.

As you can see, the best thing is to adopt a Yoga regimen **before** any of these troubles develop—for if you do, they probably **won't** develop.

BREATHING PROBLEMS: Including Asthma, Emphysema, and Chronic Bronchitis

It is popular to suppose that breathing difficulties are "irreversible." This is not so.

What has happened is that the millions of little sacs in the lungs, which pass oxygen into the blood and extract carbon dioxide from it, lose their elasticity and/or become hampered by scar tissue. Those suffering from breathing problems can find amazing relief by concentrating heavily on all the back-bending poses—Half Moon, Standing Bow Pulling, Balancing Stick, Cobra, Locust, Full Locust, Bow Pose, Fixed Firm, Camel, and the Half Tortoise are the poses that permit better and more even distribution of air in the lungs.

Even those literally gasping for breath, who are incapable of doing the whole series of exercises, should perform to the fullest extent to which they are capable the following poses: Standing Deep Breathing, concentrating on exhalation; Half Moon and Hands to Feet poses; any other backbending position they feel up to; Blowing in Firm Pose.

Your lungs are not hopeless. They respond gratefully to any exercise you deign to give them. So begin today.

BROKEN BONES

The nutrients circulated in the system when you do Yoga will certainly add to the final healing process, but the speed with which you can return to Yoga depends, of course, upon **which** bones were broken. As with postsurgical patients, have your doctor's okay, know that the bone is safely knit before you begin, and then proceed, under the supervision of a qualified teacher.

COLDS AND FLU

Do only standing postures, concentrating extra long and hard on the Balancing Stick and Triangle to increase your heartbeat as much as possible.

CONSTIPATION

Do not perform the Stretching Pose (and hence, full sit-ups) while suffering marked constipation. Concentrate instead upon the Head to Knee Pose, which is excellent for loosening things up.

DIARRHEA

Do not perform the Head to Knee Pose. Instead work extra hard and long at the Stretching Pose, which has a compressing and compacting action.

DIGESTIVE DISORDERS: Including Flatulence, Chronic Indigestion, Colitis, Ulcers, Hyperacidity

Flatulence—gas in the digestive tract—and cramping are the major features of many chronic digestive tract disorders and disease. But while the Wind Removing and Spine Twisting poses are specifically beneficial for relieving gas that is already there, the point, of course, is to prevent gas from forming in the first place.

Flatulence and hyperacidity result from overabundant ingestion of meat, starch, and processed foods (the TV dinner and greasy hamburger syndrome). Those bad eating habits are aggravated in modern society by almost total lack of exercise. This allows the food to ferment in the digestive tract, causing acidity and gas. Needless to say, Yoga moves things on their way by encouraging healthy peristalsis before they have time to ferment.

Ulcers and colitis often have a nervous or emotional basis. But again, lack of exercise, and thus poor performance on the part of the organs that should be seeing to healthy digestion and elimination, aggravates the situation. Add to this an unhealthy spine, which means a poorly functioning nervous system, and your stomach eats itself up and your intestines run wild.

As you can see, curing digestive disorders all boils down to regular yogic exercise and sensible eating habits. You might also try chewing your food instead of swallowing it whole like a shark. And ten snacks a day are better than three monstrous meals. The stomach should never be left empty, for then the powerful stomach juices have nothing to eat except the stomach lining.

GLANDULAR DISORDERS AND CHRONIC DISEASES

A malfunctioning gland asks that you get busy and get your spine—and thus your nervous system—back in order. It asks that you start breathing right and eating right, so that your blood can bring it proper nutrition, so that it can thus produce the hormones and fluids it was designed to produce, and in the proper amounts.

You can rid yourself of the majority of chronic diseases by doing your Yoga. Your body seeks homeostasis, perfect balance. Your chronic condition represents body imbalances. With your Yoga practice you can heal chronic disease conditions like heart disease, diabetes, and hypertension. Nervous disorders such as Parkinson's and multiple sclerosis give way to these Yoga exercises. No matter what your disease condition, you improve your body's ability to fight the disease every day you do your Yoga. A malfunctioning gland asks only to be a functioning gland. So get busy.

HEART PATIENTS

Under the supervision of a qualified instructor who has completed my Yoga teacher training program and with your doctor's okay, do all the exercises, but never hold them for more than five counts. Instead, repeat the poses more often than the normal student. For instance, do a combination of three sets of five counts, or four sets of three counts, being absolutely sure to **rest** between each set. It is also essential for heart patients to **breathe normally** during the postures.

Those postures that are cautionary for high blood pressure patients are also cautionary for heart patients. Take them very easy, perhaps only doing two sets of five counts at first. Common sense and perseverance are the key.

HEMORRHOIDS

The condition that produced the hemorrhoids will gradually be eliminated by my series of exercises. In the meantime, the Toe Stand has an amazing effect on that problem.

HERNIA

Those with a hernia should work extra hard on the Fixed Firm, the Tree, the Camel, and the Half Tortoise, for these poses strengthen the muscles that aggravate the problem. Those without hernia should work just as hard on these positions so that they will never have hernia problems in the first place.

HIGH BLOOD PRESSURE

Unstable (labile) high blood pressure responds so quickly to diligent Yoga practice that doctors sometimes doubt their instruments. (This quick response of the blood pressure is one of the most telling demonstrations of Yoga's ability to regulate and synchronize body systems.) If you are tested about a week after starting Yoga, you may see a slight rise in pressure. Don't be alarmed. By about the second week, that pressure will be normal or close to normal and will stay there as long as you maintain your Yoga regimen.

Consult your doctor, use common sense, and don't push hard in any of the poses the first three days. The poses in which high blood pressure patients must continue to exercise caution until their blood pressure checks out normally are these: the backward-bending portion of the Half Moon, the Standing Bow Pulling, Balancing Stick, Cobra, the third part of the Locust, Full Locust, and the Camel.

Depending on the severity of your condition, the above should be done for a count of no more than five at first, building to ten counts only after two weeks. If you are supple enough to do the Fixed Firm fully the first few days, limit that to five counts as well. As for the Bow Pose (on the floor), **a beginner with high blood pressure must never perform the pose without a qualified teacher present.**

It is because these backward-bending positions create pressure in the chest, and so on the heart, that high blood pressure patients must use caution. Do not eliminate them though—with the exception of the Bow Pose. They are the very friends you need to control your ailment.

INSOMNIA

Insomnia is caused by emotional or physical problems and imbalances, either recog-

nized or unrecognized. Once into your Yoga regimen, these imbalances will become balanced, so your body will know very well when it wants to sleep and for how long.

KIDNEY AND INTERNAL ORGAN DISORDER

Once again, perform the whole series religiously, working extra hard and long on those poses listed in Benefits as being excellent for kidney function and abdominal organs. Your body can and will right itself.

MENSTRUATION

Women vary in their menstrual reaction to Yoga. I have some students who get very bad cramps, and they take **two** classes the first day of their periods, finding that it helps them get over their cramps faster. Most women just come to the normal one class and go away feeling all better. Once in a while a woman may feel dizzy or just so lacking in strength that she can't do Yoga. If you are one of the unlucky ones, just take a little vacation. We will see you again in a few days.

MIGRAINES AND HEADACHES

Most simple headaches are caused by muscle spasms of the neck and scalp, arising from tension in the spine-back-shoulder complex. See the reference to Back Problems, do my twenty-six poses regularly, and these headaches will never bother you again.

Migraines are caused by spasms and abnormal dilations of the arteries in the brain. The full series of poses can often relieve migraine symptoms.

POSTOPERATIVE YOGA

Always be sure your doctor or surgeon has given the nod to exercise. And use common sense during any surgical recovery. If the surgery has been of the chest, abdomen, or for hernia, confine yourself to the **minimum** stretch. Only when the incision is fully healed may you work back up to full stretching. Surgery that has involved reconstruction or replacement will present a variety of problems.

Those of you recovering from spinal surgery need Yoga desperately if you are to have a healthy spine again. But you also need to do Yoga with your doctor's consent and under the supervision of a qualified instructor who has completed my Yoga teacher training program.

PREGNANCY AND POSTPARTUM

If you have been doing Yoga regularly just prior to your pregnancy, then you may continue with the Yoga exercises in this book until your third month, or up to the time you are no longer able to lie on your abdomen. Then you should use **Rajashree's Yoga for Pregnancy** video.

If you are new to Yoga or have not been doing Yoga just prior to your pregnancy, you should wait three months until you start. At that time you should use the **Rajashree's Yoga for Pregnancy** video.

If your delivery was healthy and normal, start your Yoga the moment you are out of bed. Do all the exercises from the third day, no problem.

PULLED MUSCLES OR TENDONS

Those with pulled muscles and tendons should do their Yoga. But be careful not to reinjure yourself. Perform the stretch only up to the point of the first little pain, then hold it there. The surest and fastest way to heal pulled muscles and tendons is to increase the circulation of blood to the affected parts so that they can repair themselves.

SCIATIC PAIN

The sciatic nerves begin at the junction of the lower spine, run down to the buttocks, along the outside back of the legs and down to the heel. They are the most important nerves connecting the lower part of your body with the upper part, and they are two of the largest nerves in the body. Unfortunately, the muscles associated with these nerves don't "remember" the flexibility they learn. Like strong springs or rubber bands, they go right back to Position One the moment you stop stretching them. The longer they are allowed to remain in Position One,

the more difficult it is to stretch them again. There is no alternative if one is to have a flexible and healthy sciatic system, except to exercise the sciatic nerves every single day. Athletes and ballet dancers know this, so they practice unstintingly and wear leg warmers when resting.

TENNIS ELBOW

For some strange reason, of all the students I have taught, only the Chinese and Japanese seem to have naturally flexible and thus healthy elbow joints. The rest of us run into difficulties when we engage constantly in sports or activities that put a strain on our elbows. The answer is to improve their strength and flexibility. This is done by concentrating on those positions that make the arms and elbows work in what people think of as an "unnatural" way. The best poses for Tennis Elbow are the Half Moon, Eagle, Bow Pose and Standing Bow Pulling, Balancing Stick, Locust, and Spine Twisting. This is not to be taken as a prescription to only do these postures. You have a problem and you must do all twenty-six poses to solve your problem completely. So don't shortchange yourself and seek only a half solution for your problem.

VARICOSE VEINS

Varicose veins are caused by faulty valves that allow blood to pool in the legs, stretching and enlarging the veins. Overweight, long standing, a sedentary existence, pregnancy, lack of exercise—all these things can aggravate the condition.

Many times I have seen yogic exercise return the veins to healthy working order.

Many postures work directly on the affected area, usually the legs. However, you must do all twenty-six postures, even those that do not seem to work directly on the affected area. You will find that you are using your legs one way or another in most postures. Therefore, you must perform each posture with equal vigor to assure your recovery. You will see improvement in an amazingly short time.

Bikram Yoga Schools Information

[For a listing of Affiliated Bikram Yoga
Schools in the United Kingdom and
worldwide please go to our website at
www.bikramyoga.com or contact our
headquarters at:

Bikram's Yoga College of India
International Headquarters
1862 South La Cienega Boulevard
Los Angeles, California 90035
(310) 854-5800
Fax (310) 854-6200
E-mail: info@bikramyoga.com

Make
www.thorsonselement.com
your online sanctuary